£13.39
Eng Tech 1B

Basic Engineering
Technology

YEOVIL COLLEGE
LIBRARY

TO BE RETURNED OR RENEWED BY THE DATE
LAST STAMPED BELOW

NBD to 17.5.96	2 8 JAN 2003	
27. JUN 96	1 0 FEB 2004	
19. NOV 96		
07. MAY 97	- 8 JUL 2004	
14. MAY 1997	2 9 NOV 2005	
	- 4 MAY 2006	
30. MAR 98	0 7 NOV 2006	
01. DEC 98	- 3 NOV 2009	
- 4 JAN 1999	2 2 FEB 2010	
26. JAN 99	1 5 MAR 2010	
2 5 MAY 2000	2 2 FEB 2012	
1 5 JUN 2000		
- 7 MAY 2002		
1 7 JUN 2002		
- 8 JAN 2003		

D0786065

Basic Engineering Technology

Edited by
R L TIMINGS

YEOVIL COLLEGE
LIBRARY

BUTTERWORTH
HEINEMANN

Newnes
An imprint of Butterworth-Heinemann Ltd
Linacre House, Jordan Hill, Oxford OX2 8DP

℞ A member of the Reed Elsevier plc group

OXFORD LONDON BOSTON
MUNICH NEW DELHI SINGAPORE SYDNEY
TOKYO TORONTO WELLINGTON

First published 1988
Reprinted 1989, 1990, 1991, 1992, 1993, 1995

© Butterworth-Heinemann Ltd 1988

All rights reserved. No part of this publication
may be reproduced in any material form (including
photocopying or storing in any medium by electronic
means and whether or not transiently or incidentally
to some other use of this publication) without the
written permission of the copyright holder except in
accordance with the provisions of the Copyright,
Designs and Patents Act 1988 or under the terms of a
licence issued by the Copyright Licensing Agency Ltd,
90 Tottenham Court Road, London, England W1P 9HE.
Applications for the copyright holder's written permission
to reproduce any part of this publication should be addressed
to the publishers

British Library Cataloguing in Publication Data
Timings, R. L. (Roger Leslie), 1972–
    Basic engineering technology
    1. Workshop practice
    I. Title
    670.42′3

ISBN 0 7506 0383 6

Printed and bound in Great Britain by Hartnolls Limited,
Bodmin, Cornwall

Y0008788

Contents

Preface

Basic Engineering Technology covers all the general engineering topics required for City and Guilds Basic Engineering Competences 201/1987.

The book has been planned as the successor to *Basic Engineering Craft Studies 01*, edited by Bourbousson and Ashworth (Butterworth), which was widely used and valued on the earlier City and Guilds 200 courses. The aim has been to provide, within one book, concise explanations and clear diagrams that cover the complete syllabus for the student preparing for the theory section (201-1-01) of the General Engineering qualification. Thus the text reflects the new emphasis on transferable skills, giving a broadly based introduction to vocational engineering studies.

It is hoped that the book will find a welcome not only in colleges of further education in the UK and overseas, but also in YTS training centres and in industrial apprentice training schemes.

Acknowledgement

Extracts from BS 308: Part 1: 1984 are reproduced by permission of the British Standards Institution. Complete copies of the Standard can be obtained from the BSI at Linford Wood, Milton Keynes MK14 6LE.

1

Industrial studies

1.1 The Industrial Revolution

Britain was the first country to undergo the change
from a largely agricultural economy to full-scale
industrialization. The foundations for this Industrial Revolution were laid in the seventeenth
century by the expansion of trade, the accumulation of capital, and social and political change.
This was followed in the eighteenth century by a
period of great discoveries and inventions in the
fields of materials, transportation, power sources
(steam) and mechanization. The momentum for
change originated with the mechanization of the
spinning and weaving industries which, in turn,
resulted in these cottage industries being replaced
by the factory system. This increased the need for
machines and for power units that would drive the
machines more reliably than water wheels, which
came to a halt during periods of drought. The
engineering industry was born out of this demand
for machines and engines of ever-increasing size
and sophistication.

1.2 The development of the engineering industry

The changes that began with the Industrial Revolution have continued. Some of the major developments have been:

(a) The change from an agricultural to an industrial economy

(b) The mechanization and automation of industry

(c) The development of new power sources –
steam engines, internal combustion engines,
steam and gas turbines, and electric motors

(d) The development and exploitation of natural
resources – mining for fossil fuels and metallic
ores, and the extraction and refinement of
petrochemicals

(e) The emergence of new crafts, skills and pro-
fessions in engineering, such as in electronics,
computing and aerospace.

These changes did not come about overnight, but
have steadily evolved over the centuries; some key
dates are listed in Table 1.1. In more recent times
there have been far-reaching developments in the
fields of:

(a) Standardization on a national and international scale, which has reduced cost,
improved performance and reliability and
opened up world markets

(b) Mass production and automation, which has
reduced costs and improved repeatability of
quality

(c) Communication methods such as radio, television and communication satellite links,
which have encouraged international understanding, travel and trade

(d) Microtechnology in the form of transistors,
integrated circuits and microprocessors,

which have made possible many other technologies.

Structure of the engineering industry

The main branches of industry can be summarized broadly as shown in Figure 1.1. Across all these fields of engineering there is a need for maintenance engineering and servicing if complex machines and plant systems are to work efficiently and reliably so as to give an adequate return on the capital invested in them. Further, maintenance engineering and servicing are equally important in ensuring that the machinery and plant systems work safely and do not pollute the environment.

Companies

At the start of the Industrial Revolution most of the industries were owned by private landlords and entrepreneurs, who provided the capital and often the technical inspiration. However, the rapid developments outlined above led to expansion on a scale generally beyond the resources of the individual. Nowadays we have:

Private companies   These are generally small firms where the company is wholly owned by a single person or family.

Public limited companies   The capital for a public company is subscribed by a large number of persons, insurance companies, pension funds etc., in return for shares in the company. In return for the use of this capital, the company pays the shareholders a dividend out of any profits which may be made. If the company prospers the shares may become more valuable and the individual shareholders benefit from capital growth.

Nationalized industries   These are companies owned by the state because of their strategic and social importance to the nation.

Monopolies   These companies are free from competition as they are the only companies in that particular market. This is because of the specialized nature of the product or service they offer or the high cost of capital investment involved. An example is British Gas. A group of companies acting together to reduce competition and to keep prices artificially high is called a *cartel*.

Co-operatives   These are companies owned by the workforce, management and often by customers as well. They are registered under the Industrial and Provident Societies Act. For all practical purposes they act as limited companies. However, the capital to run the company is raised by the issue of shares to the members (workers, managers and, in the retail trade, customers). The idea is to eliminate the profit element demanded

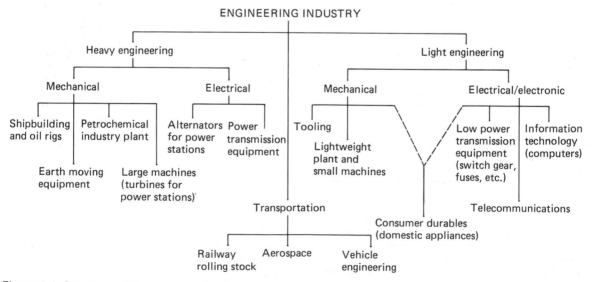

Figure 1.1 Structure of the engineering industry

Table 1.1 Some key dates in the evolution of the engineering industry

1708	The smelting of iron using coke (coal) in place of charcoal	Abraham Darby I
1708	The first practical steam pumping engine for mines	Thomas Newcomen
1750	The use of coke ovens to produce coke for iron smelting	Abraham Darby II
1769	The first spinning machine (driven by water power)	Richard Arkwright
1779	The introduction of sand moulds for casting iron	Abraham Darby III
1781	The first rotary steam engine for driving machinery	James Watt
1784	The reverberatory furnace for making better quality wrought iron	Henry Cort
1785	The first steam-powered spinning mill	Arkwright/Watt
1797	The hydraulic press	Joseph Bramah
1800	The sliding, surfacing and screw cutting centre lathe	Henry Maudslay
1808	First demonstration of a steam locomotive	Richard Trevithick
1814	First practical locomotive (the *Blucher*)	George Stephenson
1818	Planing machine for metal	Joseph Clement
1821	First commercial steam railway (Stockton to Darlington)	George Stephenson
1829	The *Rocket* wins the Rainhill steam locomotive trials and establishes many of the principles of modern locomotive design and construction	Robert Stephenson
1829	The first bench micrometer for fine measurement	
1830–60	The development of metal-cutting machine tools as they would be recognized today (lathes, drilling machines, planing, shaping and slotting machines, gear-cutting machines)	Maudslay, Nasmyth, Whitworth and others
1831	Discovery of the laws of electromagnetic induction	Michael Faraday
1839	Steam hammer for forging	James Nasmyth
1844	A dynamo for generating electricity for electroplating was produced to a patent filed in 1842: the first commercial electrical machine	Woolrich
1850	Adaptation of the vernier principle to the vernier caliper	Brown and Sharpe (USA)
1851	Adaptation of the micrometer principle to the micrometer caliper	Brown and Sharpe (USA)
1856	The use of the Bessemer converter to produce mild steel	Henry Bessemer
1860	Standardization of screw threads	Joseph Whitworth
1861	The Siemens open-hearth regenerative furnace to produce high-quality steels	William Siemens
1872	The first practical gas engine (internal combustion)	Nicholas Otto (Germany)
1886	The first practical petrol engine	Gottlieb Daimler (Germany)
1888	The first practical motor car	Karl Benz (Germany)
1889	Discovery of radio waves as forecast by Maxwell	Hughes and Hertz
1893	First commercial steam turbine	Charles Parsons
1901	First transatlantic radio transmission	Marconi (Italy)
1904	Thermionic diode valve (rectification)	Alexander Fleming
1906	Thermionic triode valve (amplification and oscillation)	Lee de Forest (USA)
1907	The first low-cost popular car for the mass market	Henry Ford (USA)
1908	The first flight by a power-driven aircraft	Wright Brothers (USA)
1912	The first production line (mass production)	Henry Ford (USA)
1944	The first electronic computer	Harvard University (USA)
1947	The transistor	J. Bardeen and W. H. Brattain (USA)
Post 1950	The microcomputer	
	Computer-controlled machine tools	
	Industrial robots	
	Space exploration	
	Use of lasers for accurate measurement	
	Development of polymeric materials and high-strength adhesives	

The dates and events listed are the first recorded *commercial applications* of the technology. In each case, many false starts and experimental prototypes were made before these dates.

by the more usual sources of capital finance; such profit is retained for reinvestment in the business after an agreed dividend has been paid to members.

1.3 Trade unions

The accumulation of vast wealth into the hands of the landowners and later the factory owners led, in many cases, to exploitation of the working class, bad working conditions and social deprivation. The trade union movement grew up to combat bad working conditions and low wages. At first legislation was passed to forbid the setting up of unions (the Combination Acts of 1799 and 1800), as the ruling classes, who dominated Parliament, were fearful of possible loss of power and influence and of the workers banding together at a time when social revolution was rife on the continent of Europe.

These Acts were repealed in 1824, and in 1825 a government measure gave trade unions the bare right to exist. Except for an amending measure in 1859, which allowed for peaceful picketing, this legislation remained in force until 1871. It was then replaced by the Trade Union Act and the Criminal Law Amendment Act, which imposed limits on strike action and swept away the right of peaceful picketing. The latter Act was repealed in 1875 and replaced by the Conspiracy and Protection of Property Act, which reinstated the right to strike and the right to picket peacefully. Various legislation has been enacted at intervals up to the present day and has alternately extended and curtailed the rights and powers of the unions according to the political complexion of the government of the day. Individual unions still fight for their members' working conditions and wages both locally and nationally.

The Trades Union Congress (TUC) originated in Manchester in 1868 and at first dominated the unions in the northern industrial districts. Separate conferences of amalgamated unions continued until 1871, when they were dissolved and the TUC took over the responsibility of coordinating the unions' fight for better working conditions. Like the unions, the TUC has been subjected to changing legislation. The structure of the TUC remained unchanged until, following the big strikes of 1919–20, the general council of the TUC was called upon to organize the general strike of 1926. Following the strike's defeat, the general council tried to adopt a more conciliatory approach to industrial relations (Mond-Turner conversations) by establishing regular consultative machinery with the employers' organizations. This was not successful. During the Second World War the government itself set up various consultative bodies concerned with industrial production, and the TUC and employers' leaders were equally represented. Although the TUC is now generally recognized as representing wide-ranging working-class interests, its authority over the unions is not mandatory and it can only bring moral pressure to bear. Some unions do not belong to the TUC.

1.4 Employer's organizations

With the growing power of the unions a counterforce was required by the employers. In 1870 a National Federation of Associated Employers was founded, and from 1890 there was a marked growth in powerful combinations. By 1936 there were 1820 employers' organizations registered and dealing with wage bargaining and labour questions in general. They also created a forum for the exchange of ideas on technical development and formed pressure groups to encourage favourable legislation.

Nowadays employers' organizations tend to fall into two categories:

(a) Those concerned with common interests in a particular trade and technology. For example, the Engineering Employers' Federation in London coordinates the engineering employers' associations, which operate at a local level and are largely concerned with wage bargaining and representing local engineering employers' interests both locally and nationally.

(b) General groupings of employers such as the Confederation of British Industry, which balances the role of the TUC and also acts as a pressure group to encourage favourable government response to the requirements of British industry at home and abroad.

In addition there are technical development associations which provide a forum for technical exchange and which carry out fundamental research and testing on behalf of the member companies: for example, the Copper Development Association (CDA) and the Motor Industry Research Association (MIRA).

1.5 Basic commercial concepts

Creation of wealth

Wealth is created by production which satisfies needs. Production, in this sense, means not only the manufacture of washing machines, cars, aircraft etc. but also the production of food from the land, the extraction of mineral and petrochemical reserves from the ground, and the provision of services such as transport, banking and insurance. Others also contribute to the creation of national wealth. For example, teachers educate the next generation of producers so that they can compete successfully in the world economic system; doctors and other health service workers keep the nation's workforce fit and active; the armed services keep open our trade routes and protect the nation's interests worldwide. Thus it can be seen that the creation of wealth is a co-operative national effort requiring the skills and talents of the whole community in the production of goods and services which can be exchanged and to which value is added at each step in the chain.

Supply and demand

In this book we are concerned with engineering products and services, not only for the engineering industry but also for the public at home and abroad. Production is governed by the economic law of supply and demand. It does not matter how good a product is, or how efficiently and cheaply it is produced: it is of no value and will create no wealth if no one wishes to buy it or if the demand is so small that the costs of production are not covered. It is the purpose of the marketing experts to anticipate the changing trends in customer needs and to ensure that design and production (supply) satisfy the demand. They may also manipulate demand through clever advertising to increase the demand for a particular product at the expense of a product made by a competing firm.

Cost

Not only must the right product be produced at the right time to satisfy demand; it must also be produced at a cost which will result in a price at which there will be sufficient customer demand to justify its production. There needs to be a balance between price and demand. Figure 1.2 shows how the cost of a product is built up. It can be seen that many factors contribute to the cost of a product, and it is essential to keep all these costs as low as possible so that the product is priced at a level the market will accept whilst producing sufficient profit to finance the continuing trading and development of the company. The overhead expenses of a company (also called the oncosts) require particular attention; they can easily get out of hand and produce a top-heavy structure. They should be kept to the minimum required to adequately service the production departments.

Suppose you draw money from you building society and start up a small business. Unless the profit you make on your investment in the business over a certain period is greater than the interest you would have got from the building society over the same period, there is no purpose in continuing with your business. If the money to start the business had to be borrowed from a bank, then there must be not only sufficient profit to finance and develop the business but also sufficient surplus to pay back the bank borrowing and the interest over an agreed period. Thus the cost of capital is an important item in the pricing of a product, no matter whether you borrow the money from yourself or from others. Large companies borrow capital on the stock market to finance their equipment and trading requirements. The dividends they pay to their shareholders may be less than the interest that could be obtained in a building society or savings bank, but the shareholder may be satisfied because the value of his or her shareholding may increase in value (capital

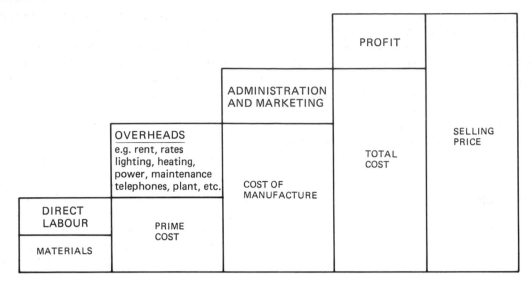

Figure 1.2 Cost structure

gains), whilst the value of money kept in a savings bank will remain static.

Inflation

Over-stimulation of the economy by an increase in money supply leads to an increase in demand for goods and services. This increase in demand in turn results in rising prices as suppliers take advantage of wider profit margins (a sellers' market) and the purchasing power of money falls. This fall in the purchasing power of money, resulting from inflation, can have an important influence on pricing and profits. Since inflation causes the value of money to diminish, each time a product is sold its price must reflect not only the current cost of production but also the fact that its replacement will cost more. The material used must be replaced at a later date with more costly material, and the labour used in production is likely to demand a rise in pay to keep pace with increases in the cost of living. If the price of the previous product has not allowed for these increases, then it may not be possible to pay for the labour and materials for the next product. Similarly, when a machine wears out the cost of the replacement will be much greater, and the retained profit must allow for replacement at this increased cost. Unfortunately, having to make these allowances frequently raises the price to a

level where there ceases to be a demand for a product or it becomes cheaper to import the product from abroad where inflation may be lower and pricing policies may be more stable. This leads to the running down of industry and an increase in unemployment in producing countries with high rates of inflation. Thus there is a need to keep inflation at or below the inflation level of other industrial nations.

Environment

Whilst manufacturing and extractive production is essential for the economic well-being of a country, it must not be allowed to destroy the environment and quality of life of the community. It is difficult to maintain a balance between the needs of industry and the needs of the environment, but great strides have been made in this direction in recent years as the importance of the environment has become more widely recognized. Disused quarries and open-cast mines are being filled in and landscaped, factories are being sited so that they are hidden by the contours of the land or by screens of trees, and greater control is being exercised over the effluent discharged from industrial plants.

The fundamental factors affecting the commercial viability of a product may be summarized as:

(a) The right product at the right price
(b) Quality: the ability of the product to operate satisfactorily for a reasonable time, to the satisfaction of the customer, so that it gives value for money
(c) Time: the need to deliver on time, and for that time to be less than that taken by competitors
(d) After-sales service, to maintain customer loyalty so that they will purchase the next generation of products.

1.6 The engineering industry

Typical products and services provided by the engineering industry are as follows:

(a) General components (nuts, bolts, washers, dowels, pipe fittings)
(b) Subassemblies (oil and fuel pumps for engines, coolant pumps for machine tools, printed circuit boards for computers)
(c) Assemblies (complete engines, machine tools, computers)

(d) Spare parts (bearings, exhaust systems, belts, clutch plates, brake shoes)
(e) Precision components and instruments (jigs, fixtures and gauges, precision measuring instruments such as micrometers, verniers, dial gauges, voltmeters, ammeters)
(f) Fabrications (ships, oil rigs, consumer durables such as cars, washing machines, hi-fi, television sets)
(g) Maintenance and repair of components and assemblies such as those listed above.

These products and services, of which the above contain but a very few examples, are provided by individuals or teams working together in laboratories, in factories or on site, using both traditional techniques and the latest advanced technologies.

1.7 Structure of a company

Figure 1.3 shows the organization chart for a typical engineering company. The roles of the main departments are as follows:

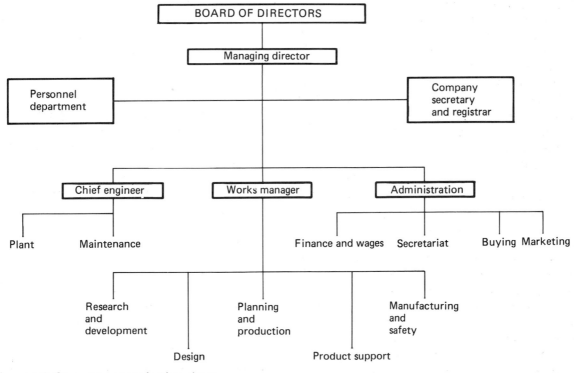

Figure 1.3 Company organization chart

Research and development This department is concerned with the development of new products and services to satisfy future customer requirements as identified by the marketing and sales department. It is also concerned with investigating new developments in technology, materials and manufacturing techniques to find new products and improve existing products whilst, if possible, reducing costs.

Design This department is concerned with product design and with the design of tooling for production. This involves not only issuing detail and assembly drawings to the shop floor, but also styling the product so that it looks attractive to the purchaser. The latter function is particularly important in the case of consumer durables.

Testing This department is most important in maintaining quality. It tests not only the finished product, but also samples of incoming materials, part-finished components and subassemblies to ensure that the company's suppliers are maintaining standards.

Planning and production This department is concerned with planning the manufacturing programme. It must ensure that the right product is produced at the right time and that the plant and workforce are kept fully employed with a uniform work load, and it must organize the smooth running of the production line.

Manufacturing This department works closely with planning and production in the actual manufacture of the product in the workshops. It is also concerned with providing the necessary plant and equipment.

Marketing and sales This is one of the most important departments. Unless the product can be sold there is no point in producing it. Marketing is concerned with determining where the markets exist for the products being made and what the future requirements of the markets will be. They must not only predict when the markets for particular products are coming to an end so that the company is not left with surplus stock, but also ensure that new and suitable products are available for new market opportunities. The sales force can then act on this market information to sell the company's products.

Product support This department is concerned with three areas of work:

(a) After-sales service and commissioning to ensure customer satisfaction.
(b) Sales engineering. For example, in the machine tool industry, skilled engineers who are experts in the uses and applications of the company's products can back up the sales force by advice and assistance to the customer in the correct and most profitable use of the equipment purchased.
(c) Training the customer's workforce in the use of the plant and equipment purchased.

Commercial administration It has already been stated that the purpose of production is the creation of wealth through the exploitation of the company's products. Therefore the commercial departments are crucial, since they include personnel, training and clerical services and, most important of all, the financial control of the company.

1.8 Personnel in a company

Management personnel

There are several levels of management in a company. If it is a private or a public limited company there will be:

Board of directors The directors are elected to the board because they are persons of great technical and commercial experience. They are usually professionally qualified in law, accountancy, banking, insurance, technology and marketing. Their responsibility is to plan the strategy of the company, to ensure that the strategy is competently executed, to sanction large-scale capital expenditure, and to represent the shareholders' interests and ensure that their invested capital is used wisely. In a public company the shareholders have the ultimate power to vote individual directors or even the whole board out of office if they are not satisfied with the running of the company, although this rarely happens. The directors communicate with executive management of the company via the managing director.

Departmental managers are responsible to the managing director for the implementation of company policy in their departments and for the organization and efficient running of their departments. They are expected to show initiative in solving problems and to feed suggestions for the more efficient running of the company back to the board via the managing director.

Supervisors are responsible to the departmental managers for the efficient running of their sections within the department. They supervise the workforce directly to ensure that it is employed profitably and that the policy of the department is implemented efficiently.

Technical personnel

There are several categories of technical personnel employed in engineering. The functions and career patterns of these are as follows:

Chartered engineers These are technologists who hold a first degree in engineering and have a number of years' experience in their chosen branch of the engineering profession in a responsible position. Some older chartered engineers may have academic qualifications other than a first or higher degree.

Technical engineers usually have Higher National Certificates or Diplomas and even degrees in engineering. In a large company they will work under the direction of a chartered engineer and, in turn, will have technicians and craftsmen working for them. Some technical engineers may themselves become chartered engineers when they have served sufficient years in a suitably responsible position. In smaller companies the technical engineer is often the most senior engineer in the plant.

Engineering technicians usually have an Ordinary National Certificate or equivalent qualification. They work under the direction of a chartered engineer or a technical engineer and, in turn, may have craftspersons and operatives working for them. As well as proven technical ability, a company will be looking for initiative and leadership qualities, as many technicians soon become section leaders and enter management as they continue to improve their qualifications.

Craftspersons will have served a craft apprenticeship and will have considerable manual skills

and practical experience plus appropriate technical qualifications such as those provided by the City and Guilds of London Institute. Again, a company will be looking for initiative and leadership qualities as well as proven practical ability, as many craftspersons move on to become chargehands and foremen. After further study and suitable bridging qualifications they may also become technicians in research and development and in work study where their background of practical experience is of great importance. In small companies some may even become managers, but larger companies tend to recruit management trainees from the ranks of the technical engineers and graduate engineers.

Operatives will usually have undergone limited skill training with or without some basic technical education. Where suitable personal qualities are shown, they may be trained to become setters and chargehands of production sections. A few may take further qualifications and move further up the ladder, but this is not usual.

1.9 Education and training

The differences in academic qualifications and training of the various categories of technical personnel have already been introduced. These qualifications can be gained at:

Universities and polytechnics First degrees and diplomas, higher degrees and research degrees.

Polytechnics and colleges of further education Ordinary and Higher National Certificates and Diplomas of the Business and Technician Education Council (BTEC).

Colleges of further education Craft and operative certificates of the City and Guilds of London Institute, and special category certificates of that institution at a much higher level (e.g. various specialisms within computer-aided engineering).

Practical skill training at all levels – craft, technician and management – is organized through universities, polytechnics, colleges and the training departments of industry by such bodies as:

Engineering Industry Training Board, which supervises the practical skill training elements at all levels within the engineering industry on a national basis. It sets standards of achievement to be attained and advises companies and colleges on the content and organization of training programmes to ensure uniformity.

Manpower Services Commission (MSC), which encourages, approves and sponsors adult training and retraining courses in co-operation with employers and local education authorities. It is also responsible for the YTS training schemes.

These courses of study and training may be full time, sandwich (partly in company, partly in college or university), part time, or at the place of employment. In addition there is the Open University through which technical degrees may be obtained by spare-time study at home.

Further reading

On the history of engineering see:

Guest, G. M. *A Brief History of Engineering*
Merril, J. M. *Arkwright of Crompton*
Steeds, W. *A History of Machine Tools* 1700–1910

On requirements for the engineering profession see:

Engineering Council *Standards and Routes to Registration*

Exercises

For each exercise, select *one* of the four alternatives.

1 At the start of the Industrial Revolution most industries were owned by
   (a) the nation
   (b) stockbrokers
   (c) private landlords and entrepreneurs
   (d) public institutions
2 In public companies the shares
   (a) are owned by a private individual or family
   (b) are owned by a nationalized industry board
   (c) are quoted on the stock market
   (d) can be bought direct from the company by the general public.

3 The Combination Acts of 1799 and 1800
   (a) forbade the setting up of trade unions
   (b) made the setting up of unions legal for the first time
   (c) recommended the control of individual unions by the TUC
   (d) allowed trade unions to be set up on a limited scale.
4 Employers' organizations (such as the Engineering Employers' Federation) grew up
   (a) solely to exchange technical information
   (b) as a counterforce to the growing power of the unions
   (c) to assist in developing export trade
   (d) to standardize components.
5 It is a basic economic concept that
   (a) production is good irrespective of need
   (b) the creation of wealth is unrelated to production
   (c) production, in this sense, refers only to manufactured goods such as cars
   (d) wealth is created by production which satisfies needs.
6 Supply and demand and cost are economic concepts:
   (a) they are closely related
   (b) they are totally unrelated
   (c) only supply and demand are related
   (d) only demand and cost are related.
7 Inflation, in the economic sense
   (a) increases competitiveness
   (b) increases purchasing power
   (c) reduces purchasing power
   (d) reduces the cost of living
8 The cost of manufacture refers to
   (a) material and labour costs only
   (b) labour and overhead costs only
   (c) prime costs only
   (d) prime and overhead costs combined.
9 Figure 1.4 shows
   (a) a cost structure for a company
   (b) an organization chart for a company
   (c) the structure of the engineering industry
   (d) the organization chart of an engineering employers' organization.
10 The person primarily responsible for communication between the board of directors and

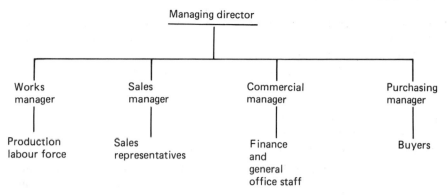

Figure 1.4

the departmental managers in a limited company is the
(a) managing director
(b) chairman of the board
(c) company secretary
(d) personnel director.

11 A person holding an honours degree in a branch of engineering and who has a number of years' experience in a responsible position in that branch of engineering is likely to be
(a) a technical engineer
(b) a chartered engineer
(c) an engineering technician
(d) a craftsperson.

12 An engineering craftsperson is likely to hold a certificate of the City and Guilds of London Institute after studying at a
(a) university
(b) polytechnic
(c) college of further education
(d) sixth-form college.

2

Observing safe practices

2.1 Health and safety legislation

The Health and Safety at Work etc. Act 1974 provides a comprehensive and integrated system of law dealing with the health, safety and welfare of people at work, and also of the general public as affected by work activities. There are six main provisions:

(a) Creation of a Health and Safety Commission. This consists of three trade union members appointed by the TUC, three management members appointed by the CBI, two local authority members and one independent member representing the general public. There is a full-time, independent chairperson. They have taken over the responsibility previously held by various departments for the control of most occupational safety and health matters. The Commission is also responsible for the Health and Safety Executive.
(b) Reorganization and unification of the various government inspectorates, such as the Factory Inspectorate and the Mines and Quarries Inspectorate, into a single Health and Safety Executive (HSE). The inspectors of the HSE have wider powers than previously and their duties are to enforce the policies of the Commission (see Section 2.2).
(c) Modernization and overhaul of all existing laws dealing with safety, health and welfare at work.
(d) Making employers responsible for maintaining a safe place to work and consulting on safety matters with their employees.
(e) Provision of new powers and penalties for the enforcement of safety laws (see Section 2.2).
(f) Establishment of new methods of accident prevention, and new ways of operating future safety regulations.

The Health and Safety at Work Act is superimposed upon the existing legislation of 31 relevant Acts and some 500 subsidiary regulations, and these remain in force until they are replaced by new legislation. Examples of such Acts are:

Factories Act
Offices, Shops and Railway Premises Act.

Examples of subsidiary regulations are:

Abrasive Wheel Regulations 1970
Milling Machines Regulations 1972
Protection of Eyes Regulations 1974
Safety Representatives and Safety Committees Regulations 1977
Notification of Accidents and Dangerous Occurrences Regulations 1980
Health and Safety (First-Aid) Regulations 1981.

2.2 Health and Safety Executive

The inspectors of the Health and Safety Executive have three lines of action they may take if they find contraventions of new or existing legislation:

Improvement notice If there is a legal contravention of any of the relevant statutory provisions the inspector issues an improvement notice. This requires the fault to be remedied within a specified time, but work may continue in the meantime.

Prohibition notice If there is a risk of personal injury, the inspector can issue a prohibition notice. This immediately stops the activity until it has been made safe to the inspector's satisfaction. In addition, the inspector can seize, render harmless or destroy any substance or article the inspector considers to be the cause of imminent danger or serious personal injury.

Prosecution In addition to the above actions, the inspector can prosecute any person (employer, employee or supplier) contravening a relevant statutory provision.

2.3 Employers' responsibilities

Safe place to work For example, the premises must not represent a fire hazard. There must be easy access and exit. Stairways and floors must be safe. The premises must be properly heated and ventilated. There must be adequate lighting, and proper toilet facilities.

Safe plant and equipment For example, machinery must be regularly maintained and fitted with proper guards. Boilers and compressed air receivers must be regularly inspected and tested. Lifting tackle must be regularly inspected and tested and clearly marked with the safe working load. Electrical equipment and installations must adhere strictly to the requirements of the IEE Regulations and any special regulations for particular processes.

Safe system of work In addition to the plant and equipment being safe, there must be a safe method of working, a code of safe practice and protective clothing when needed. For example, there must be: the observance of strict no-smoking rules when handling flammable substances; the isolation of electrical circuits before working on them; the wearing of respirators and protective clothing when cleaning out chemical storage vats; and the use of goggles and face shields with the correct filter glasses when welding.

Safe working environment The need for a properly heated and ventilated environment with adequate sanitation, washing and first-aid facilities has already been introduced as part of a safe place of work. In addition, some processes give off harmful and/or toxic fumes and there must be proper extraction for these from the working environment. However, under the Health and Safety at Work Act such fumes must be rendered harmless before being released to the atmosphere so that they do not become a nuisance or a hazard to the general public. Dust must also be controlled, especially the dust from grinding and polishing processes. An unpleasant environment can cause accidents due to fatigue and carelessness.

Safe methods of handling, storing and transporting goods Materials in store must be safely racked up or stacked so that they cannot slip and fall whilst in store or being removed. Adequate lifting and trucking facilities must be provided so that materials and work in progress can be handled safely, and staff training must be provided in the use of such equipment and facilities. Chemicals must be kept in properly marked containers away from foodstuffs. Gas cylinders and flammable substances must be kept in ventilated compounds away from the working areas and protected from accidental damage. They must be shielded from direct sunlight and frost.

Reporting accidents and incidents A register of all accidents, however slight, must be kept in the workplace. Certain types of accidents and dangerous occurrences must be reported by the employer to the Factory Inspectorate of the Health and Safety Executive. The inspectors have the right to investigate any incident and to inspect the register on demand.

Information, instruction, training and supervision It is the responsibility of the employer to ensure that all employees are properly trained in the use of equipment and in codes of practice to ensure they work in a safe manner. Training is particularly important when new processes are introduced. There is a legal obligation to inform staff if a hazardous material is to be used and to train them in its proper use and in emergency procedures. Young and inexperienced workers must be continuously supervised when working on potentially dangerous machines or processes.

Safety policy   An employer with more than six employees must produce a written safety policy. The policy must be made available to every employee and it should be regularly reviewed by a joint committee of management and employees' representatives. Where trade unions operate within a company, they are allowed to appoint their own safety representatives.

2.4 Employees' responsibilities

Under the Health and Safety at Work Act the employee can also be prosecuted for breaking the safety laws. The employee is legally bound to co-operate with the employer to enable the employer to comply with the requirements of the Act.

Personal health and safety   It is the legal responsibility of all workers to take reasonable care of their own health and safety, and to ensure that they act in a responsible manner so as not to endanger other workers or members of the public.

Misuse of equipment   It is an offence under the Act to misuse or interfere with equipment provided for your health and safety or the health and safety of others.

2.5 Causes of accidents

Human carelessness   Most accidents are caused by human carelessness. This can range from slipshod attitudes and the deliberate disregard of safety regulations and codes of practice to improper behaviour and dress. Carelessness can also result from fatigue and ill-health caused by lack of attention to the working environment.

Personal habits   Personal habits such as alcohol and drug abuse can render a worker a hazard not only to himself but also to other workers. Fatigue due to second-job 'moonlighting' can also be a considerable hazard, particularly when operating machines.

Supervision and training   As processes become increasingly complex it is essential that all operators are given adequate training not only in the process but also in emergency procedures. It is also essential that adequate supervision is provided to ensure that the correct procedures are carried out and also to prevent human carelessness and poor personal habits.

Environment   Unguarded and badly maintained plant and equipment is an obvious cause of injury, but the most common accidents are due to falls on slippery floors, poorly maintained stairways and scaffolding, and crowded workplaces with obstructed passageways. Bad lighting, inadequate ventilation and noise can lead to headaches, fatigue and carelessness. Dirty surroundings can also lead to a lowering of personal hygiene standards.

2.6 Accident prevention

Elimination of the hazard   This is essential in preventing accidents. Where necessary the workplace should be tidied up with clearly defined passageways, improved lighting and ventilation, reduced noise and non-slip floorings.

Removal of the hazard   Hazardous processes should be replaced with something less dangerous. For example, asbestos brake and clutch linings should be substituted by linings made of safer, synthetic materials.

Guarding   Rotating machinery, drive belts and cutters must be securely fenced or guarded so that the operators cannot come into contact with such equipment.

Personal protection   Approved overalls, safety helmets, safety shoes, earmuffs, respirators and eye protection must be worn singly or in combination depending upon the process and safety regulations in force. Such protective clothing and appliances must be provided by the employer when the process demands their use.

Safety education   This is important in producing positive attitudes towards safe working practices and habits. Warning notices and instructional posters should be displayed in prominent positions. Information, education and training should be provided in all aspects of health and safety: for instance, process training, personal hygiene, first aid and fire procedures.

2.7 Personal attitudes to safety

It is essential that all employees adopt a positive attitude towards safety, not only for their own welfare but also to ensure the safety of other employees and the general public. Therefore employees must make a positive decision to act and work in a careful and responsible manner, understand all the regulations and codes of practice, adhere to them rigidly, and never indulge in horseplay and foolish behaviour in the workplace. To this end, employees need to know: that dangers can occur; what protection is available, and how to use it; and how to prevent accidents.

2.8 Safety procedures

Personal hygiene Before work it is preferable to rub a barrier cream into your hands. This fills the pores of the skin with a water-soluble antiseptic cream, so that when you wash your hands the dirt and germs are removed with the cream. It also prevents dirt penetrating the pores of your skin in the first place. Always wash at the end of a work period, before and after using the toilet, and before handling food. Never use solvents for cleaning your hands, as these can attack the skin. Change overalls regularly and before they become too dirty and a health hazard.

Protective clothing This should be worn wherever necessary. Figure 2.1 contrasts satisfactory and unsatisfactory protective clothing. Overalls should be in good condition and close fitting so that they will not become entangled with moving machinery. Eye protection must also be worn when performing such operations as grinding, machining and welding. Safety boots or shoes should also be used; these have steel toecaps to prevent crushing, and steel innersoles to protect the foot in the event of stepping on sharp objects.

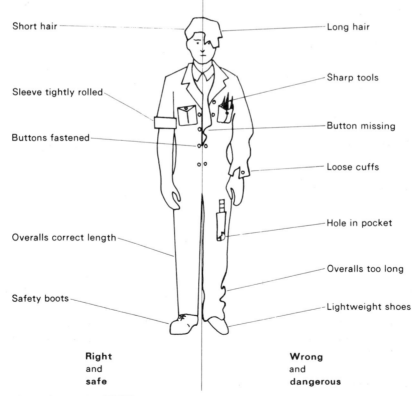

Short hair — Long hair
Sleeve tightly rolled — Sharp tools
Buttons fastened — Button missing
— Loose cuffs
Overalls correct length — Hole in pocket
— Overalls too long
Safety boots — Lightweight shoes

Right and **safe**   **Wrong** and **dangerous**

Figure 2.1 Correct dress (courtesy EITB)

(a) WEAR THE CORRECT TYPE OF PROTECTIVE CLOTHING

(b) PROTECT THE HEAD

(c) WEAR SAFETY FOOTWEAR

Figure 2.2 **Additional protection**

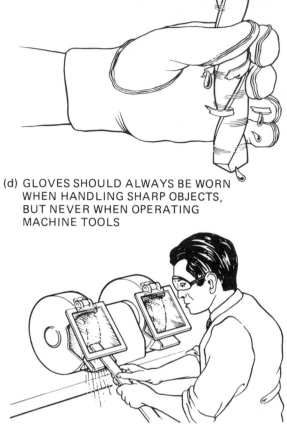

(d) GLOVES SHOULD ALWAYS BE WORN WHEN HANDLING SHARP OBJECTS, BUT NEVER WHEN OPERATING MACHINE TOOLS

(e) ALWAYS PROTECT THE EYES WHEN USING MACHINERY

(f) WEAR A SUITABLE RESPIRATOR WHEN DUST AND FUMES ARE PRESENT

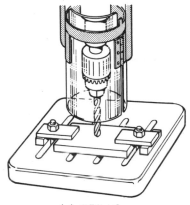

(a) DRILLS

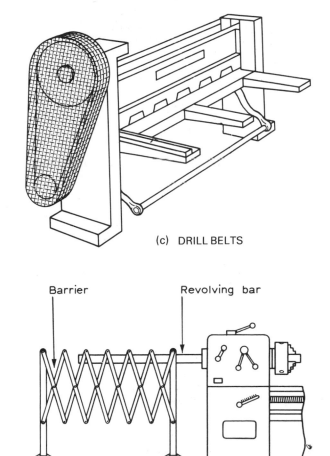

(c) DRILL BELTS

(b) CUTTERS

Barrier   Revolving bar

(d) PROTRUDING PARTS

Figure 2.3 Guards

Industrial gloves should be used when handling sharp objects and sheet metal. Safety helmets should be worn on erection sites and where overhead cranes are in use. Respirators should be used for protection against dust and fumes, and earmuffs should be used for protection against excessive noise. Examples of this additional protection are shown in Figure 2.2.

Guards   Gearboxes, drive belts, revolving shafts and cutters must be guarded by law to prevent accidental contact. Removal of such guards or interference with them is a contravention of the Health and Safety at Work Act and can lead to prosecution – assuming the operative concerned survives any accident resulting from such foolish actions. Some examples of guarding are shown in Figure 2.3.

2.9 Electrical hazards: legislation and regulations

Electrical equipment is potentially dangerous. The hazards associated with it can be summarized as:

(a) Electric shock
(b) Fire due to the overheating of cables, connections and appliances
(c) Explosions due to using unsuitable equipment when flammable vapours are present.

These hazards can result from badly designed installations and equipment, inadequate maintenance, deliberate overloading of installations and equipment, tampering with equipment by un-trained and unauthorized personnel, and the selection of unsuitable equipment for a particular application.

To reduce the risks of electrical accidents, the following laws and codes of practice must be observed:

Health and Safety at Work etc. Act 1974
Electricity (Factories Act) Special Regulations 1908 and 1944
Institution of Electrical Engineers' 'Regulations for the Electrical Equipment of Buildings'.

Unlike the Acts of Parliament, the IEE Regulations are not mandatory. However, they are used by all qualified and reputable installation and maintenance engineers since their adoption ensures a satisfactory standard of safe practice.

2.10 Electrical hazards: general safety rules

Personal safety

Before using any electrical equipment it is advisable to carry out a number of visual checks (Figure 2.4):

(a) Check that the cable is not damaged or frayed.
(b) Check that the cable is properly secured at both ends and that none of the conductors is visible.

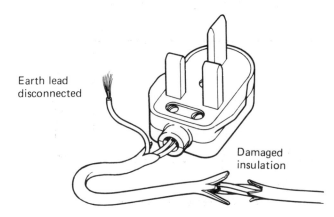

Earth lead disconnected

Damaged insulation

Figure 2.4 Examine cables and plugs daily

(c) Check that the plug is in good condition.
(d) Check that portable power tools are suitable for working from a reduced voltage (110 V or 55 V) and that a suitable transformer is supplied.
(e) Check that even where a low-voltage transformer is used, the equipment is connected to the mains via an earth-leakage circuit breaker (ELCB: see later).
(f) Check that metal-clad equipment has a properly connected earth lead (earth continuity conductor) and is fitted with a properly connected three-pin plug.

Figure 2.5 shows the proper connections to a standard three-pin plug. As well as ensuring that the wires are correctly connected according to their colour code, check that the fuse has the correct current rating (the minimum required for the correct operation of the connected appliance). Further references to the colour coding of electrical cables may be found in Table 9.2.

When working with electricity always make certain that the circuit is isolated from the supply by switching it off and/or removing the fuses before touching any of the conductors. Make sure that the circuit cannot be turned on whilst you are working on it. Either lock the switch in the off position and keep the key in your pocket, or draw the fuses and keep these in your pocket.

Check that all circuits are dead by means of a reliable voltage indicator or a test lamp before commencing work on the conductors.

Ensure first-class workmanship and proper design on all jobs, no matter how insignificant they may seem. Always work to the recommendations of the IEE Regulations.

Earthing

One of the conductors in the supply system is connected to earth. This is done for reasons of safety to persons and property. For instance, if a fault occurred which caused a live conductor to come into direct contact with the metal frame of an electrical appliance, current would flow through the frame to earth via any available path. Such a fault current would cause fires and even death.

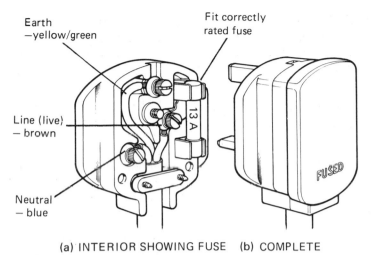

(a) INTERIOR SHOWING FUSE   (b) COMPLETE

Figure 2.5 Correctly connected plug with 13 A fuse

Figure 2.6 shows how a person may receive an electric shock. The human body is able to act as a conductor of electricity and therefore receive a shock. The amount of current passing through the body will depend upon the applied voltage and the resistance of the body (which varies from person to person). A person may receive a shock by touching the live conductor of the supply whilst being in contact with earth (Figure 2.6a). A person may receive a shock by touching the live conductor of the supply and the neutral conductor at the same time (Figure 2.6b).

For example, consider the portable metal-clad electric drill shown in Figure 2.7a. It shows a fault in which the live cable has come into contact with the metal casing. A person holding the drill would complete the circuit to earth through his or her body and would receive a severe electric shock. Figure 2.7b shows how this potentially lethal condition can be avoided. All metal-clad appliances should be connected to earth. On portable appliances this is done by ensuring that the green and yellow striped conductor of a three-core cable is connected between the metal frame of the appliance and the earth pin of the plug (see Figure 2.5). In the event of the fault shown, the fault current has alternative paths. Since the resistance of the human body is very much greater than that of the earth wire the current will flow through the

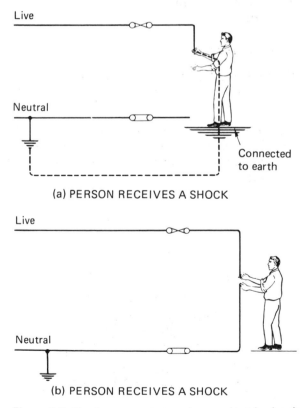

(a) PERSON RECEIVES A SHOCK

(b) PERSON RECEIVES A SHOCK

Figure 2.6 The two ways to receive an electric shock

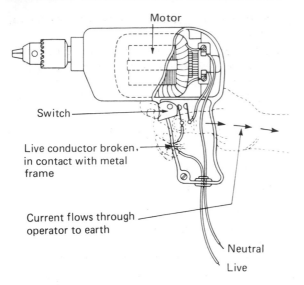

(a) ELECTRIC DRILL NOT EARTHED

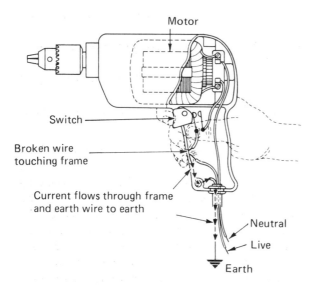

(b) ELECTRIL DRILL IS EARTHED

Figure 2.7 Need for earthing on an electric drill

earth wire and, if it is of sufficient magnitude, it will blow the fuse in the plug and isolate the circuit. However, small fault currents may flow undetected. For this reason, portable appliances should be further protected by using a low-voltage supply

and by being fed through an earth-leakage circuit breaker (ELCB). This device detects very small currents flowing to earth and turns off the supply as soon as such earth-leakage currents flow.

Alternatively, portable appliances and power tools are increasingly designed and manufactured using double-insulated techniques, so that earthing is not required. In addition to the insulation around the individual conductors, the whole assembly is built into an insulated container. For example, modern portable power drills are housed in moulded plastic bodies instead of cast metal bodies. If there is any exposed metalwork this is isolated from any electrical components by the design of the plastic casing. Double insulation is to be preferred since there is always the possibility of the earth conductor failing by incorrect connection, accidental disconnection, or the conductor breaking.

2.11 Procedure in case of electric shock

(a) Switch off the supply of electricity, if this can be done quickly.
(b) Do *not* touch the person's body with your bare hands. Human flesh is a conductor and you would receive a severe shock.
(c) Remove the affected person from contact with the supply using insulating material such as dry clothing, a dry sack, or any dry plastic material which may be handy.
(d) If the injured person is not breathing or there is any doubt, commence artificial respiration *immediately*. When a person is not breathing the brain is starved of oxygen and permanent damage may occur within a few minutes.
(e) Send for medical help.

2.12 Artificial respiration

Use the method with which you are most familiar and continue until the casualty is breathing satisfactorily or until medical help arrives.

Mouth-to-mouth resuscitation

Positioning Working from the side, hold the casualty's head in both hands as shown in Figure 2.8. One hand presses the head backwards and the

Figure 2.8 Mouth-to-mouth resuscitation: positioning

other pushes the lower jaw upwards and forwards. This opens the air passage. Remove dentures and any obstruction to breathing (e.g. chewing gum).

Inflation Seal the casualty's nose by pinching the nostrils together. Take a deep breath, open your mouth wide and place it round the mouth of the casualty and blow until you see the casualty's chest rise.

Deflation Turn your head away and allow the casualty's chest to fall whilst you take another breath. Repeat the cycle of inflation and deflation at about 12 cycles per minute until the casualty commences to breathe satisfactorily, or until expert medical help arrives and you are asked to cease resuscitation.

2.13 Fire prevention, procedures and legislation

Fire

Fire or burning is the rapid combination of a fuel with oxygen (air) at high temperature. A fire can reach a temperature of 1000°C within a few minutes of starting. For a fire to start there are three requirements:

(a) Combustible (flammable) material (the fuel).

(b) Sufficient oxygen (air) to sustain the oxidation (burning) process.

(c) Sufficient heat to raise the oxygen/fuel mixture to a suitable temperature for the combination of the fuel with the oxygen (the burning) to commence. Burning is an exothermic reaction (heat is given out), and sufficient heat may be generated not only to maintain the fire but also to spread it.

There are many causes of outbreaks of fire. Some of the more common are:

(a) Careless and illicit smoking
(b) Process heating getting out of control
(c) Accidents with gas torches
(d) Faults developing in heating equipment
(e) Spontaneous combustion (e.g. oily overalls or rags can ignite spontaneously at room temperature in an oxygen-rich atmosphere, and care must therefore be taken when using oxy-acetylene equipment in confined places)
(f) Fuel leaks
(g) Mechanical faults causing overheating of machinery due to friction
(h) Inadequate ventilation allowing a build-up of flammable solvent vapours
(i) Electrical faults in installations and appliances.

Fires can spread rapidly. Once established, even a small fire can generate sufficient heat energy to spread and accelerate the fire to surrounding combustible materials. The fire can spread by the radiation of heat to surrounding objects, by the conduction of heat along steel girders, and by the convection of hot gases up life shafts and staircases. Convection is the most common cause of fire spread, and for this reason doors and windows should be kept shut.

Fire prevention

Fire prevention is largely 'good housekeeping'. The workplace should be kept clean and tidy. Oily rags and waste materials should be disposed of in the metal bins provided. The bins should have airtight lids to smother any fire which starts. Plant, machinery, furnaces and space heating equipment should be regularly inspected and maintained.

Electrical installations, alterations and repairs must only be carried out by qualified personnel and must be to the standards laid down by law and by the IEE Regulations. Smoking must be banned where there is any risk from flammable substances and, in any case, should only be allowed in supervised areas with adequate ashtrays provided and regularly emptied. Flammable materials should be stored in properly designed compounds away from the working areas, and only the minimum required for the shift should be brought into the workplace. Advice should be sought from the fire prevention officer of the local fire brigade on the precautions which should be taken and on any processes involving heating and the use of flammable substances. This advice should be regularly updated, particularly when new processes and materials are introduced.

Fire procedures

Fire procedures are as follows:

(a) Raise the alarm and call the fire service. Large factories often have a local fire-fighting team on site.

(b) Evacuate the premises. Regular fire drills must be held and all personnel must be familiar with normal and alternative escape routes. There must be assembly points and a roll call of personnel. A designated person must be assigned to each floor or department to ensure that evacuation is complete, and there must be a central reporting point.

(c) Keep fire doors closed to prevent the spread of smoke, which causes panic, particularly on stairways. Emergency exits must be kept unlocked whilst the premises are in use, and lifts must not be used during a fire.

(d) Finally, attempt to contain the fire until the local or public brigade arrives, unless the spread of the fire or the generation of poisonous fumes from burning plastic materials makes this dangerous, in which case evacuate the scene of the fire immediately. Saving life is more important than saving property.

Fire protection legislation

The main purpose of such legislation is to safe-guard lives rather than property. General standards of fire safety are governed by:

Health and Safety at Work Act 1974
Building Regulations 1976
Fire Precautions Act 1971
Fire Services Act 1947

There is additional legislation covering specific risks such as public places of entertainment, the storage and use of highly flammable gases, petrol and petroleum products, the storage and use of explosives, and so on.

In addition, employers are required by law to:

(a) Obtain a fire certificate for premises where more than 20 persons are employed (more than 10 if not at ground level)

(b) Ensure the premises have unobstructed escape routes and that these are clearly indicated and known to all employees

(c) Provide adequate and suitable fire extinguishers and alarms

(d) Hold regular fire drills which must be logged along with the time taken to evacuate the premises

(e) Give training in the use of fire-fighting equipment, the use of alarms and the procedures for summoning the public fire service

(f) Inspect and test fire alarms and fire-fighting equipment regularly, and log the findings of such inspections and tests and any remedial action that has to be taken.

2.14 Fire-fighting equipment

As described in Section 2.13, a fire is caused by a combination of three factors: a combustible material; a sufficiently high temperature; and oxygen (air). Successful fire fighting therefore depends upon the removal of one or more of these factors.

Water extinguishers

Water works mainly by cooling down the burning substances, but also by smothering the fire with the steam produced. A fine spray of water is more effective than a concentrated jet, since the force of the latter may scatter the burning material and actually spread the fire. However, a powerful jet is

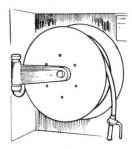

(a) HOSE REEL

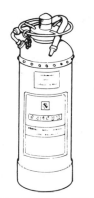

(b) PRESSURIZED WATER
EXTINGUISHER

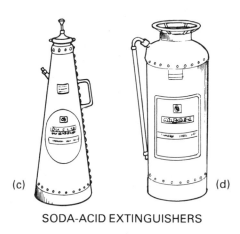

(c)   (d)

SODA-ACID EXTINGUISHERS

Figure 2.9 Water extinguishers

and releases the high-pressure gas which forces out the water.

Soda-acid extinguisher (Figure 2.9c) With this type, striking the plunger causes an acid to mix with sodium carbonate solution. The reaction generates high-pressure carbon dioxide which forces out water under pressure.

Soda-acid extinguisher (Figure 2.9d) This type is operated by inverting the extinguisher to bring the reactive solutions into contact.

Water extinguishers are for use on ordinary combustible materials such as wood, paper and textiles.

Carbon dioxide extinguisher

Carbon dioxide (CO_2) is a gas in which most substances will not burn. It works by displacing the oxygen surrounding the fire. The extinguisher (Figure 2.10) is only suitable in confined places where the carbon dioxide gas cannot be removed by draughts before the fire is smothered and put out.

This type of extinguisher is used for fires in electrical appliances and small flammable liquid fires.

Figure 2.10 Carbon dioxide (CO_2) extinguisher

required to reach a large fire because the heat makes it impossible to approach the seat of the fire and use a spray. There are various types of water extinguisher:

Hose reel (Figure 2.9a) Check whether the water has to be turned on or whether it comes on automatically when the reel is pulled out. The hose outlet can provide either a jet or a spray as required.

Pressurized water extinguisher (Figure 2.9b) To operate, remove the cap and strike the knob. This punctures a compressed carbon dioxide cartridge

Foam extinguisher

Reactive chemicals in the extinguisher generate a foam under high pressure when the chemicals are mixed. Mixing is achieved by removing the cover cap and striking the plunger (Figure 2.11a) or by

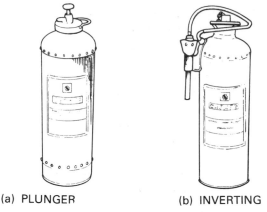

(a) PLUNGER   (b) INVERTING

Figure 2.11 Foam extinguishers

inverting the extinguisher (Figure 2.11b) according to the instructions on the body of the extinguisher. Foam works by blanketing the fire and excluding the air. It also cools the combustible material below the ignition temperature.

This type of extinguisher is used for fires involving flammable liquids such as oils, fats, solvents, petrol and paint.

Vaporizing liquid extinguisher

This contains liquids which produce a heavy vapour when released; the vapour smothers the fire. Unfortunately the vapour is highly toxic (poisonous) and the extinguisher must not be used in confined spaces. The extinguisher (Figure 2.12) is operated by pumping the handle and spraying a jet of the liquid at the seat of the fire.

Figure 2.12 Vaporizing liquid extinguisher

This type of extinguisher is for use out-of-doors on motor vehicle fires and for small fires in electrical equipment.

Dry powder extinguisher

This ejects a fine powder under high pressure at the fire. The powder (basically sodium bicarbonate) not only smothers the fire, but breaks down when heated to give off carbon dioxide which also helps to smother the fire (Figure 2.13).

The main advantage of these extinguishers is that the powder is non-toxic and is easily removed with a vacuum cleaner. Dry powder extinguishers are therefore highly suitable for use in kitchens and food stores. They can be used for all types of flammable liquids indoors and outdoors. They can also be used on metallic fires (e.g. burning magnesium).

Figure 2.13 Dry powder extinguisher

Fire blankets

These are woven from fire-resistant synthetic fibres and rely on their ability to smother the fire (Figure 2.14). The old-fashioned blankets made from asbestos must *not* be used. The blankets are pulled from the container and placed over the fire.

Fire blankets are suitable for use in the home, in workshops and in laboratories. They are also used where a person's clothing is on fire, by rolling them up in the blanket and smothering the fire. Do not cover the person's face.

Warning   Do not use water or foam extinguishers on fires in electrical equipment or where live

Figure 2.14 Fire blanket

electrical equipment is present. Such use would give the person operating the equipment a severe or even lethal shock.

2.15 General safety rules

These can be summarized as follows:

(a) Be alert to actual or potential hazards at all times.
(b) Report all hazards and potentially dangerous incidents in order to prevent an accident.
(c) Maintain a high standard of personal hygiene.
(d) Aim to protect not only yourself but other people as well.
(e) Know and observe all safety rules and emergency procedures. Delay can cost lives.

Exercises

For each exercise, select *one* of the four alternatives.

1 Under the Health and Safety at Work Act, safety in the workplace is the responsibility of
  (a) the employer alone
  (b) the employee alone
  (c) the employer and the employee jointly
  (d) an inspector from the Health and Safety Executive alone.
2 If an inspector of the Health and Safety Executive needs to stop an activity where there is a risk of personal injury, he issues
  (a) an improvement notice
  (b) a prohibition notice
  (c) a writ
  (d) a certificate.
3 Most accidents are caused by
  (a) human carelessness
  (b) unsafe stairs and ladders
  (c) faulty machines
  (d) faulty electric wiring.
4 Approved safety glasses should be worn
  (a) only when machining
  (b) only when welding
  (c) whenever there is danger to the eyes
  (d) when working in bad light.
5 The main safety requirement for overalls is that they should be
  (a) lightweight and not restrict movement
  (b) non-absorbent
  (c) fireproof
  (d) in good condition and close fitting so that they will not become entangled in machinery.
6 The main reason for providing guards on all machines and machine tools is to
  (a) prevent damage to precision mechanisms
  (b) contain the coolant
  (c) prevent any person coming into contact with dangerous and moving parts
  (d) prevent swarf (chips) from damaging the slideways.
7 The earth lead in a flexible cable is coloured
  (a) green and yellow
  (b) brown
  (c) blue
  (d) black.
8 Double-insulated portable power tools
  (a) must be earthed
  (b) do not need to be earthed
  (c) can only be used in very low-voltage supplies
  (d) are unsuitable for use on sites.
9 All metal-clad electrical equipment must be earthed to prevent
  (a) electric shock
  (b) fuses repeatedly blowing
  (c) radio interference
  (d) loss of power.

10 In the event of severe electric shock
    (a) commence artificial respiration immediately
    (b) remove the victim from contact with the supply and wait for medical help
    (c) remove the victim from contact with the supply and commence artificial respiration immediately
    (d) leave the victim alone whilst you fetch expert medical help.

11 The main reason for not using water on fires in electrical equipment is because
    (a) the equipment will be damaged
    (b) the person fighting the fire will receive an electric shock
    (c) the fuse supplying the equipment will blow
    (d) water will aggravate this type of fire.

12 The type of extinguisher which must *not* be used on fires involving flammable liquids such as oil is
    (a) foam
    (b) carbon dioxide
    (c) vaporizing liquid
    (d) pressurized water.

13 A fire certificate is required for
    (a) all industrial premises
    (b) no industrial premises
    (c) premises where more than 20 persons are employed
    (d) premises which are not at ground level where any number of persons are employed.

14 Regular fire drills must be held and
    (a) the date logged
    (b) the time taken to evacuate the premises logged
    (c) the date and the time taken to evacuate the premises logged
    (d) no log is required if fewer than 20 persons are employed.

15 Portable power tools must be supplied at low voltage from a transformer
    (a) only on site
    (b) at all times to prevent the risk of severe electric shock
    (c) to prevent damage to the equipment if overloaded
    (d) to protect the supply cables.

16 The purpose of a fuse is to
    (a) protect a circuit from excess current in the event of a fault developing
    (b) prevent an appliance connected to the circuit from breaking down
    (c) avoid the need for an isolating switch
    (d) avoid the need for earthing metal-clad equipment.

17 A barrier cream is used to
    (a) keep your skin soft
    (b) prevent dirt from entering the pores of your skin
    (c) avoid having to use soap when washing
    (d) act as an insulator and prevent electric shock.

18 It is every person's responsibility to report
    (a) only serious accidents
    (b) all hazards and potentially dangerous incidents
    (c) only incidents which affect them personally
    (d) only incidents which affect the general public.

3

Movement of loads

3.1 Loads and safety

Within the context of the engineering industry, loads are defined as heavy objects such as machines and large pieces of metal which have to be moved about within the factory. When large, heavy and often ungainly objects have to be moved, safety precautions and correct handling procedures are very important. The movement of heavy loads involves careful planning and the anticipation of potential hazards before they arise. Safety precautions and codes of practice must be strictly observed.

3.2 Manual handling

The manual handling of objects is the cause of more injuries in factories and on site than any other factor. Back strains and injuries are the main causes of absenteeism in industry. Extreme care must be taken when lifting or moving heavy or awkward objects manually. The rules for correctly lifting a load are as follows:

(a) Maintain the back straight and upright; bend the knees, and let the strong muscles of the legs and thighs do the work. Figure 3.1a shows a good lifting position.
(b) Keep the arms straight and close to the body.
(c) Balance the load, using both hands if possible.

(d) Avoid sudden movements and twisting of the spine.
(e) Take account of the position of the centre of gravity of the load when lifting.
(f) Clear all obstacles from the vicinity.
(g) Take care to avoid injury to other people, particularly when moving long loads.
(h) Use gloves or 'palms' to avoid injuries from sharp or rough edges.
(i) Take care when moving materials which are wrapped and greased.
(j) Plan the move to avoid unnecessary lifting.
(k) Move the load by the simplest method – simple is safe.
(l) Where a team of people are involved, ensure that they are of similar physique and that they work as a team, obeying only the member who is in charge.
(m) Never obstruct your vision with the load you are carrying (Figure 3.1b).

Objects which are too heavy to carry can often be easily moved by using rollers as shown in Figure 3.1c, providing that the underside of the object is free from obstructions. Suitable rollers would be thick-walled tubes or solid bars. Ensure that the rollers are of sufficient width to prevent the load toppling. Turning a corner can be effected by placing the leading roller at an angle. Always take care when the ground level rises or falls. A crowbar can be used to raise the load initially so that the rollers can be placed beneath it. Figure 3.1d shows a crowbar being used to raise a load.

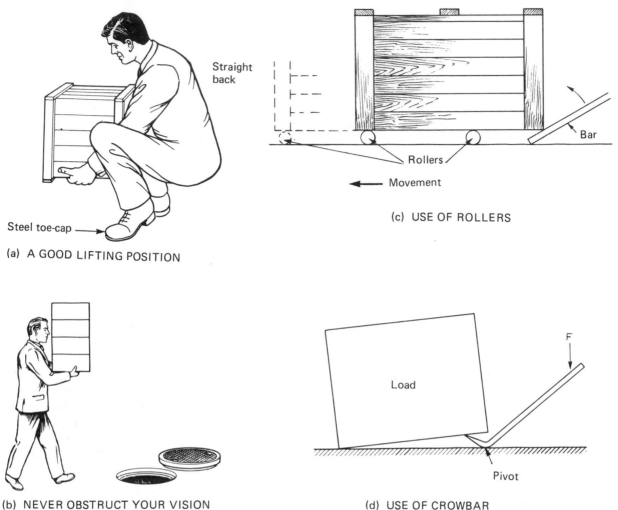

(a) A GOOD LIFTING POSITION

(b) NEVER OBSTRUCT YOUR VISION

(c) USE OF ROLLERS

(d) USE OF CROWBAR

Figure 3.1 Manual handling

3.3 Mechanical lifting gear

Lifting gear can be classified by the motive power used to operate it:

Manual (muscle power) Lifting gear powered manually includes pulley blocks and chain blocks.

Electrical (electric motor) There are many types of electrically operated hoists used in factories. These range from electrically powered chain blocks to overhead gantry cranes capable of lifting massive loads.

Pneumatic (compressed air) These are used in mines, oil refineries and chemical plants where there are flammable and explosive vapours. Electrically power equipment is not suitable under such conditions because of the possibility of sparks from the motor brushgear and from the switchgear controlling the motor. They are limited to relatively small loads.

Hydraulic (pressurized liquids) These may be direct acting using long-stroke pistons and cylinders, or they may use hydraulic motors to rotate the lifting rope drum. These systems are used where greater power is required than can be provided by compressed air. For mobile cranes a single engine and hydraulic pump can power a number of hydraulic motors, through a simple system of valves and pipes, for slewing and luffing the jib and for raising and lowering the load without the need for complicated gearboxes and drive shafts.

Petrol and diesel engines   These are used to power mobile cranes and hoists for use on building and civil engineering sites. They power the crane or hoist through mechanical or hydraulic transmission systems. (Except for simple hoists, the hydraulic transmission system is the most used nowadays.)

3.4 General safety rules

The following rules must be strictly observed when moving heavy loads, as accidents resulting from falling loads can lead to very serious or fatal injuries and cause costly damage.

(a) As a general rule, loads over 20 kg require mechanical lifting gear.

(b) The safe working load (SWL) marked on the hoist and on such accessories as slings must *never* be exceeded.

(c) Shock loading of lifting equipment must be avoided, as must swinging and twisting.

(d) The hook must be positioned above the centre of gravity of the load so as to maintain an even balance and eliminate any chance of the load tilting and slipping. Never lift by the point of the hook as shown in Figure 3.2a.

(e) Pushing or pulling the load to adjust its position should be avoided.

(f) Loads must not be transported over the heads of people working below; nor should anyone walk under the load as shown in Figure 3.2b.

(g) Loads should not be left hanging without someone in attendance. Preferably they should be lowered on to a suitable support until required.

(h) The load should always be lowered gently into position and secured so that it cannot slip before removing the lifting equipment.

(i) Stack materials safely, as shown in Figure 3.2c, so that they cannot slip.

(j) Slinging is a skilled job. The attachment of slings to the load and the crane hook should

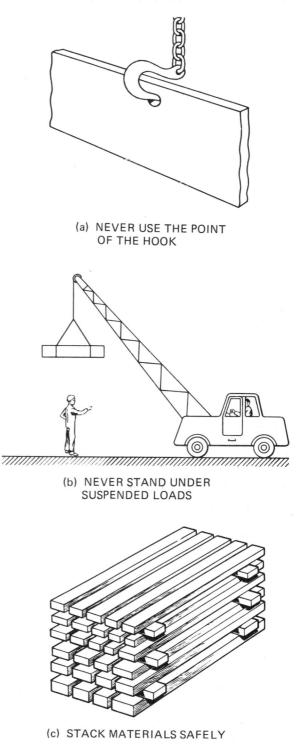

(a) NEVER USE THE POINT OF THE HOOK

(b) NEVER STAND UNDER SUSPENDED LOADS

(c) STACK MATERIALS SAFELY

Figure 3.2 General safe handling rules

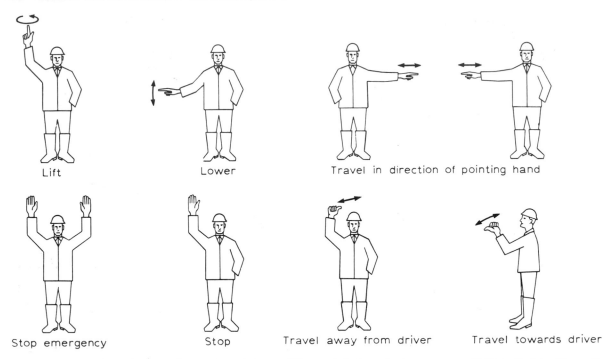

Lift   Lower   Travel in direction of pointing hand

Stop emergency   Stop   Travel away from driver   Travel towards driver

Figure 3.3 Slinger's signals (reproduced from *Slingers' Safety Code*, courtesy British Safety Council, 163/173 Praed Street, London W2, from whom copies may be obtained)

only be done by a trained and experienced slinger, who alone should give the signals to the crane driver. All signals must be given clearly and distinctly by one person only. The eight basic signals illustrated in Figure 3.3 must not be varied.

3.5 Accessories for lifting gear

Hooks

Hooks for lifting gear are made from forged steel, and are carefully proportioned to prevent the load from slipping off the hook. Hooks are frequently painted bright yellow to attract attention and prevent people walking into them.

Slings

Fibre ropes and steel wire ropes must be properly spliced at the joints. Eyes (loops) at the ends of slings must also be spliced and lined with metal thimbles to prevent chafing. Slings may be continuous as shown in Figure 3.4a, or they may have two, three or four legs as shown in Figure 3.4b.

Slings are used as follows:

(a) Two-leg slings should be long enough to avoid an excessive angle between the legs. When the angle at the top of the sling reaches 120°, the load on each leg equals the load being lifted. Figure 3.5 shows how the load on the sling increases as the top angle is increased.

(b) Slings must never be overbent, nor must they be crushed against the sharp corners of the load. Protection must be provided as shown in Figure 3.6.

(c) Worn or damaged slings as shown in Figure 3.7a must never be used.

(d) Slings must never be twisted or kinked, nor must bolts or knots be used to shorten them as shown in Figure 3.7b.

(e) Worn or damaged slings must never be used.

(f) The safe working load (SWL) marked on the sling must never be exceeded. Figure 3.8 shows the SWL for fibre and wire rope slings and chain slings.

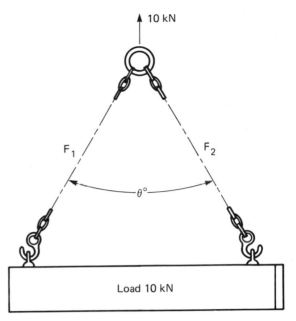

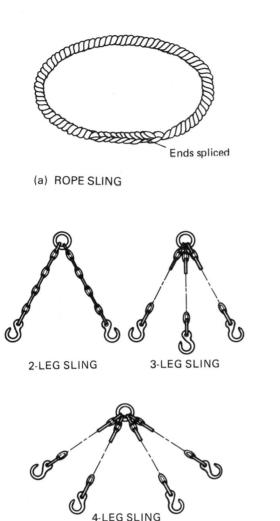

(a) ROPE SLING

2-LEG SLING   3-LEG SLING

4-LEG SLING

(b) CHAIN SLING

Figure 3.4 Slings

Angle between sling legs $\theta °$	Forces acting on legs	
	F_1	F_2
30	5.2 kN	5.2 kN
45	5.6 kN	5.6 kN
60	5.8 kN	5.8 kN
90	7.0 kN	7.0 kN
120	10.0 kN	10.0 kN
150	20.0 kN	20.0 kN
180	∞	∞

Figure 3.5 Load on sling legs

Eyebolts and shackles

Eyebolts must be designed and constructed to approved standards (BS 4278) and correctly used. They are often used for convenience in lifting engines, gearboxes and electric motors, as shown in Figure 3.9a. When used they must be tightened down to the shoulder, as shown in Figure 3.9b.

Shackles must also be designed and manufactured to approved standards (BS 3551) and correctly used. They are frequently used to connect the eye of a wire rope to an eyebolt as shown in

Figure 3.9c. Shackles are made in a variety of sizes, and it is important to match the SWL of the shackle with that of the wire rope and of the eyebolt. Remember the old adage: 'A chain is only as strong as its weakest link'.

Chains and rings

Chain slings have already been introduced. Direct attachment of the chain to the load and the crane hook can cause damage to the links of the chain. The chain should be provided with a ring for

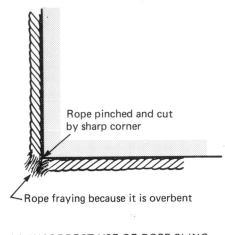

(a) INCORRECT USE OF ROPE SLING

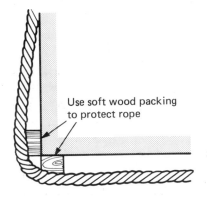

(b) CORRECT USE OF ROPE SLING

Figure 3.6 Use of rope sling

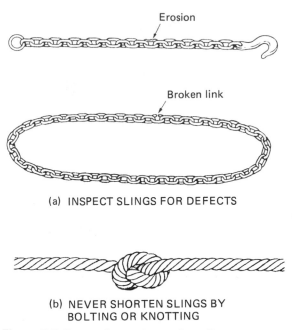

(a) INSPECT SLINGS FOR DEFECTS

(b) NEVER SHORTEN SLINGS BY
BOLTING OR KNOTTING

Figure 3.7 Precautions when using slings

attachment to the crane hook, and hooks for attachment to the load, as shown in Figures 3.4 and 3.5. Chains should be regularly inspected, as the links wear thin and even fracture in use. They are also subject to work hardening and metal fatigue. The repeated stressing of metal components such as the links in the chain causes the granular structure of the metal to become distorted. When this occurs, the metal becomes harder and more brittle and will eventually break when subjected to normally safe loads. For example, if a thin strip of metal is held in a vice and bent backward and forward repeatedly, it will work harden at the point of bending and eventually snap off. Work

hardening can be corrected by heat treatment (normalizing). The chain is heated to about red heat in a furnace and allowed to cool slowly in still air. This restores the grain structure to its normal condition.

Fibre ropes can also be damaged by direct attachment to crane hooks. The use of a ring as shown in Figure 3.10 prevents damage to the rope and also prevents slipping.

Special-purpose equipment

Special loads often warrant the manufacture of special-purpose equipment. For example, the spreader shown in Figure 3.11a may be used where wide or long loads have to be lifted regularly; it avoids having to spread a two-leg sling to a wide angle. The plate clamps shown in Figure 3.11b are another example of special-purpose lifting equipment.

Inspection

It is a legal requirement under the Health and Safety at Work Act that all lifting gear has to be regularly inspected by qualified engineers specializing in this work. The results of the inspection and

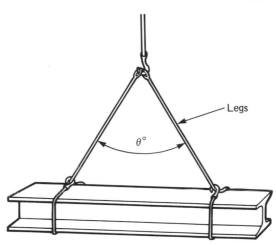

SAFE WORKING LOAD (kN)

Diameter mm	$\theta = 45°$			$\theta = 90°$			$\theta = 120°$			Single leg		
	S	C	F	S	C	F	S	C	F	S	C	F
10	14	17	—	10	13	—	7	9	—	7	9	—
12	18	25	—	14	19	—	10	13	—	10	13	—
16	37	44	3	28	35	3	20	24	2	20	24	2
20	55	70	5	42	54	4	30	38	3	30	38	3
24	74	100	7	57	78	6	40	54	4	40	54	4
28	109	136	11	84	105	8	60	74	6	60	74	6
32	147	175	15	113	134	11	80	95	8	80	95	8
36	184	220	18	140	170	14	100	120	10	100	120	10
40	220	276	—	170	210	—	120	150	—	120	150	—

S   steel wire rope of 6 × 24 × 7 construction made from wire of 1300–1400 MPa tensile strength
C   short-link chain
F   fibre rope laid up 3 × 3

Figure 3.8 Typical safe working loads (kN)

tests must be recorded in the register provided. If the inspector condemns any item of equipment it must be taken out of service and destroyed immediately. The inspector also confirms the SWL markings on the equipment as being correct. No item of lifting equipment must be taken into service until it has been inspected and certificated.

3.6 Transporting loads

Various types of truck are used in factories for moving heavy loads. Care must be taken when transporting loads, particularly when working on site. Advise the safety officer of the movement of any large or heavy object, particularly if there are special risks involved. Contact the workshop supervisor if the movement requires machines or other plant to be shut down. Take care when transporting overhanging loads through narrow passageways, especially when flanked by machines and other equipment. Watch overhead clearances. Keep clear of processes involving molten metals or dangerous chemicals. Always ensure you have adequate lighting. Take care if persons are working overhead: decorators or builders using ladders and scaffolding are particularly vulnerable.

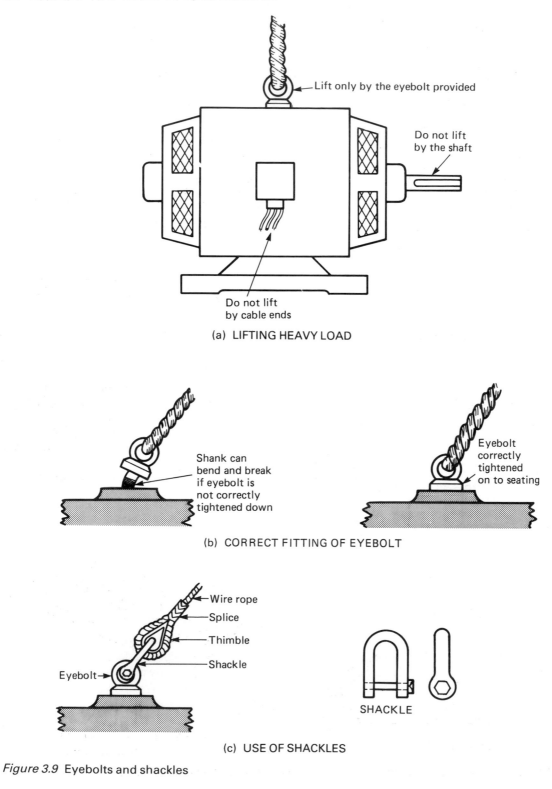

Lift only by the eyebolt provided

Do not lift by the shaft

Do not lift by cable ends

(a) LIFTING HEAVY LOAD

Shank can bend and break if eyebolt is not correctly tightened down

Eyebolt correctly tightened on to seating

(b) CORRECT FITTING OF EYEBOLT

Wire rope
Splice
Thimble
Shackle
Eyebolt

SHACKLE

(c) USE OF SHACKLES

Figure 3.9 Eyebolts and shackles

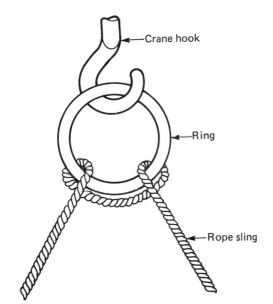

Crane hook

Ring

Rope sling

Figure 3.10 Use of ring with rope sling (for use of rings with chain slings see Figures 3.4 and 3.5)

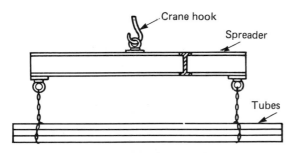

Crane hook

Spreader

Tubes

(a) SPREADER FOR LIFTING LONG LOADS

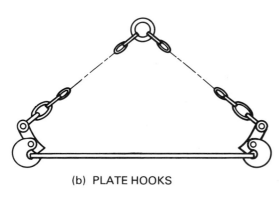

(b) PLATE HOOKS

Figure 3.11 Special purpose equipment

(a) CHECK CONDITION OF GROUND

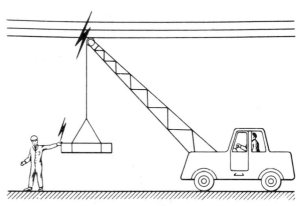

(b) BEWARE OF HIGH-VOLTAGE CABLES

(c) NEVER TAKE CHANCES
ON INCLINED SURFACES

Figure 3.12 Precautions on site

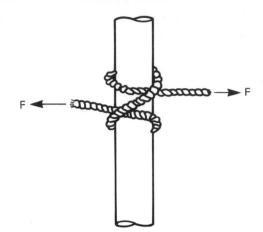

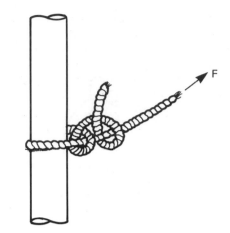

(a) CLOVE HITCH

(b) TWO HALF HITCHES

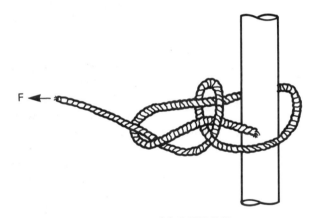

(c) BOWLINE

(d) REEF KNOT

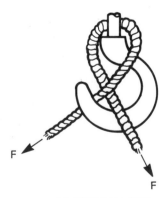

(e) SINGLE LOOP

Figure 3.13 Useful knots: F is direction of force exerted by load (pull)

Movement on site requires special care, particularly if visibility is obstructed by fog, rain or snow. Pay attention to soil conditions. Unstable ground can be very dangerous for mobile cranes, as shown in Figure 3.12a. When using cranes, also watch out for overhead electric cables (Figure 3.12b). Special care must be taken when moving heavy loads on inclined surfaces (Figure 3.12c).

3.7 Knots in ropes and slings

Various knots are used in conjunction with fibre ropes and slings. Never tie knots in wire ropes and chains; they are not sufficiently flexible and will be permanently damaged. Figure 3.13 shows some important knots:

Reef knot   For joining ropes of equal thickness.
Clove hitch   For fastening a rope to a pole or bar.
Single loop   For preventing a rope slipping from a crane hook.
Half-hitch   Two are used for securing a rope to a sling or ring.
Bowline   To form a loop which will not tighten under load.

Exercises

For each exercise, select *one* of the four alternatives.

1 When lifting loads manually
   (a) keep the back straight and upright, bend the knees and let the leg and thigh muscles do the work
   (b) keep the knees straight, bend the back and let the back muscles do the work
   (c) bend both the back and the legs and let the arm muscles do the work
   (d) keep the legs and back straight and only use the arms.
2 A load which is too heavy to lift can still be easily moved manually by
   (a) dragging it along the floor
   (b) dragging it along planks
   (c) using rollers or skates
   (d) rocking it from corner to corner.

3 As a general rule, mechanical lifting gear is required for loads over
   (a) 1 kg
   (b) 20 kg
   (c) 100 kg
   (d) 200 kg.
4 The safe working load (SWL) is
   (a) the least load for which the equipment should be used
   (b) the maximum load which should not be exceeded
   (c) an indication of the average load for which the equipment is suitable
   (d) the only load for which the equipment is suitable.
5 A crane hook should be positioned
   (a) at the centre of gravity of the load
   (b) below the centre of gravity of the load
   (c) above the centre of gravity of the load
   (d) either side of the centre of gravity of the load.
6 Heavy loads requiring cranes
   (a) should not be transported over the heads of persons working below
   (b) can be transported over persons working below if they are wearing safety helmets
   (c) can be transported over persons working below without special precautions
   (d) can be transported over persons working below providing the load does not exceed the SWL of the crane.
7 The slingers' signal illustrated in Figure 3.14 indicates
   (a) come forward
   (b) go back
   (c) raise the load
   (d) make an emergency stop.

Figure 3.14

8 Continuous slings made from fibre or wire ropes are joined at the ends
  (a) by splicing
  (b) with a reef knot
  (c) with a clamp
  (d) using an epoxy adhesive.

9 The force acting on each leg of the sling shown in Figure 3.15 is
  (a) $F_1 = F_2 = 5$ kN
  (b) $F_1 = F_2 = 10$ kN
  (c) $F_1 = F_2 = 15$ kN
  (d) $F_1 = F_2 = 20$ kN.

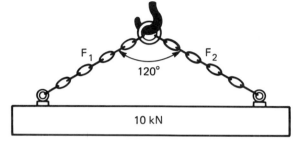

Figure 3.15

10 If the angle between the legs of the sling shown in Figure 3.15 is increased (becomes greater than 120°), the loads acting on the legs of the sling will
  (a) remain the same
  (b) be reduced
  (c) be increased
  (d) eventually become zero when the angle is increased to 180°.

11 The device shown in Figure 3.16 is
  (a) a shackle
  (b) an eyebolt
  (c) a chain ring
  (d) a non-slip crane hook.

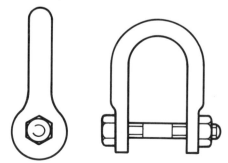

Figure 3.16

12 Figure 3.17 shows a
  (a) reef knot
  (b) clove hitch
  (c) bowline
  (d) half-hitch.

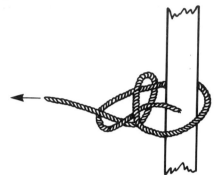

Figure 3.17

13 The knot shown in Figure 3.17 is used to
  (a) make a loop which will not tighten or slip
  (b) join two ropes
  (c) fasten a rope to a pole
  (d) make a loop which will tighten under load.

14 Knots are suitable for use with
  (a) wire ropes
  (b) fibre ropes
  (c) wire and fibre ropes
  (d) thin chains, wire ropes and fibre ropes.

15 The wooden packing shown in Figure 3.18 is used
  (a) to take up the slack in the sling
  (b) to protect the load
  (c) to prevent the load cutting into the sling
  (d) so that the load can be lifted by a fork-lift truck when it is resting on the ground.

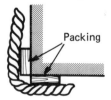

Figure 3.18

16 A shackle is used to
  (a) connect a chain ring to an eyebolt
  (b) in place of an eyebolt for heavy loads
  (c) in place of a spliced eye at the end of a fibre or wire rope
  (d) in place of a chain hook.

17 It is a legal requirement of the Health and Safety at Work Act that lifting gear must be
 (a) checked occasionally
 (b) inspected regularly by a qualified engineer
 (c) inspected only when first taken into use
 (d) inspected only if the SWL is exceeded.

4

Measurement and dimensional control

4.1 Measurement as a comparator process

Figure 4.1 shows the length of a metal bar being measured by comparing it with a steel measuring tape. Whenever measurement takes place, the object being measured is compared with some known standard – in this example, the measuring tape. Therefore measurement is a *comparator process*. The object being measured can be compared with non-indicating and indicating equipment.

Non-indicating equipment

A 'go' and 'not go' caliper gauge can be used to check a dimension to determine whether or not it

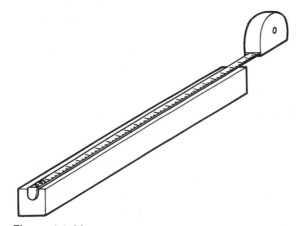

Figure 4.1 Measurement as a comparator process

lies within the limits of size set by the designer, as shown in Figure 4.2. A try-square can be used to check whether or not two surfaces are at right angles to each other. In neither of these examples can the equipment used indicate the actual thickness or the actual angle. Therefore non-indicating equipment is used for checking rather than measuring.

Indicating equipment

A rule or a steel measuring tape can be used to measure a dimension (Figure 4.1). It will determine not only whether the dimension lies within the limits of size set by the designer, but also the actual size of the dimension. A bevel protractor can be used to measure the angle between two surfaces. In both these examples the equipment used indicates the actual size of the dimension or the actual size of the angle. Therefore indicating equipment is used for measuring operations. An example of a comparator for measuring linear (length) dimensions is shown in Figure 4.14.

4.2 Dimensional properties

There are many dimensional properties which have to be considered when checking or measuring a component. The more important ones are as follows.

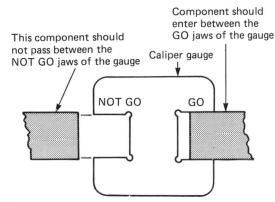

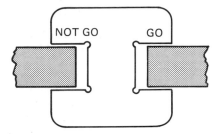

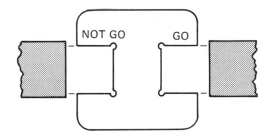

(a) CORRECTLY SIZED COMPONENT ENTERS 'GO'
    JAWS BUT NOT 'NOT GO' JAWS

(b) UNDERSIZE COMPONENT ENTERS 'GO' AND
    'NOT GO'

(c) OVERSIZE COMPONENT DOES NOT ENTER 'GO'
    OR 'NOT GO'

Figure 4.2 Use of caliper gauge

Length

Measurement of length, or linear measurement, is the measurement of the distance between two points in a straight line. Irrespective of the name given to the distance being measured (e.g. thickness, width, height, diameter) it is still a measurement of length. The main unit of measurement used in mechanical engineering is the millimetre.

(Although no longer a preferred unit, the inch is still widely used in the engineering industry.) Some techniques for the measurement of length will be considered in Section 4.5.

Flatness

Flatness is the extent to which the surface of a component deviates from a true plane. In practice the surface being checked is compared with a surface plate or a straight edge. Figures 4.3a d show a flat component being compared with a surface plate. In this example the surface plate is the standard of flatness. A little engineer's blue is lightly smeared on the surface of the plate and the workpiece is then rubbed on the surface plate with a rotary motion. High spots on the workpiece will pick up blue from the surface plate, and the flatness can be judged by the pattern of the high spots indicated. This technique presupposes that the surface plate is itself flat. Figure 4.3e shows how three surface plates worked together will ensure almost perfect flatness. Two plates A and B will appear to be flat when checked against each other using engineer's blue. Plates A and C will also appear to be flat, since their surfaces are in overall contact. However, comparison of plates B and C shows this not to be true. The plates are then scraped until they show themselves to be flat when compared in any combination. All three plates can only show overall contact if they are perfectly flat. Do not confuse the engineer's blue used in conjunction with surface plates with the blue lacquer used for marking out.

Figure 4.4 shows a straight edge being used to check flatness. Since the straight edge will sit across the two highest points, any hollows can be judged by the light gap between the blade of the straight edge and the component.

Parallelism

This is the constancy of distance between two planes – that is, lack of taper. Figure 4.5 shows a shaft being checked for parallelism by comparing its diameter at each end with calipers after taking an initial cut. This is particularly important when turning between centres. The tailstock should be adjusted to remove any errors of parallelism (see Section 6.5).

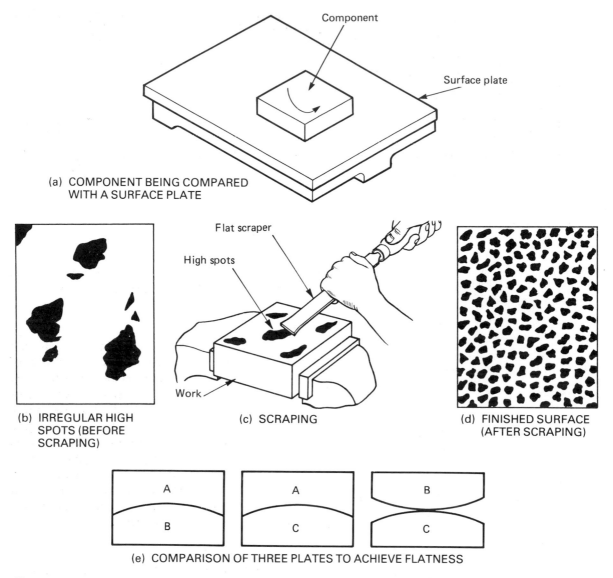

(a) COMPONENT BEING COMPARED WITH A SURFACE PLATE

(b) IRREGULAR HIGH SPOTS (BEFORE SCRAPING)

(c) SCRAPING

(d) FINISHED SURFACE (AFTER SCRAPING)

(e) COMPARISON OF THREE PLATES TO ACHIEVE FLATNESS

Figure 4.3 Flatness

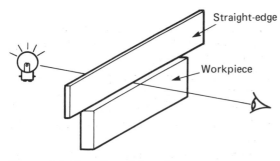

Figure 4.4 Use of straight edge

Surface roughness

Surface roughness or surface texture is the deviation of the surface of a component from a perfectly true and smooth plane. Figure 4.6a shows the difference between roughness and waviness. Roughness is the irregularities in the surface texture which are inherent in the production process: for example, the tooling marks left when turning on the lathe with a coarse feed. Waviness is

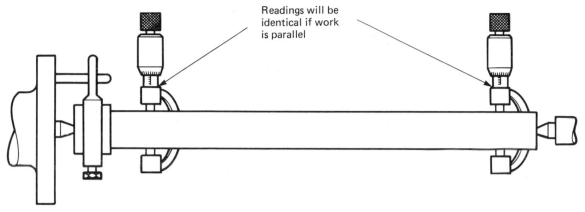

Readings will be
identical if work
is parallel

Figure 4.5 Parallelism

a slow undulation in the component surface upon which the roughness is superimposed. It may result from such factors as machine and work deflections, slideway wear, vibrations and chatter. When assessing roughness, a mean line is drawn through the recorded profile as shown in Figure 4.6b, so that the areas above the line are equal to the areas below the line over a distance called the sampling length (*l*). The most commonly used international measure of roughness is the *Ra* value. This is the arithmetic mean of the variation in the roughness profile from the mean line measured in micrometres (μm).

Surface roughness can be assessed by using comparison blocks. These come in boxed sets and represent typical surface roughness values as produced by typical workshop processes. Comparison is made by drawing a fingernail across the component and then across the test blocks until a match in feel is obtained. This requires much skill and experience and is of very limited accuracy. The more precise way of assessing roughness or texture is the use of an electronic measuring machine. A stylus and pickup head (similar to that in a record player) is drawn across the workpiece; deviations in the path of the stylus are converted into electrical impulses, amplified and used to plot a magnified replica of the surface on a paper tape.

Angles

An angle is a measure of the inclination of one line to another, or of a surface to a reference plane.

There are many ways of measuring angles, and some of these will be considered in Section 4.6.

Profiles

The profile is the outline of the component, and usually consists of straight lines and radii (arcs of circles). These may be checked by templates or by projection of the component on to a screen. Templates may consist of a standard screw-cutting centre gauge (Figure 4.7a), radius gauges (Figure 4.7b) or special templates for turned components (Figure 4.7c). The principle of a typical optical projector is shown in Figure 4.7d. It throws an enlarged image of the component on to the screen where it can be compared with an enlarged template or a master transparency.

Relative position

The alignment of large components and subassemblies prior to final assembly is accomplished in a variety of ways:

(a) Horizontal and vertical planes can be set using precision bubble levels (spirit levels), as shown in Figure 4.8a.
(b) Alignment in the vertical plane can also be achieved using a plumb-bob, as shown in Figure 4.8b. A plumb-bob always hangs vertically: that is, it points towards the centre of the earth.
(c) A light source and sights may be used as shown in Figure 4.8c.

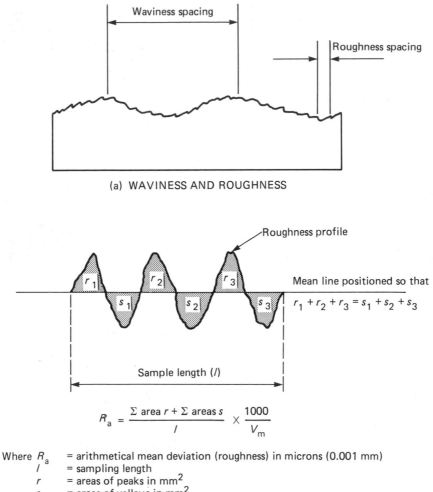

(a) WAVINESS AND ROUGHNESS

$$R_a = \frac{\Sigma \text{ area } r + \Sigma \text{ areas } s}{l} \times \frac{1000}{V_m}$$

Where R_a = arithmetical mean deviation (roughness) in microns (0.001 mm)
      l = sampling length
      r = areas of peaks in mm^2
      s = areas of valleys in mm^2
      V_m = vertical magnification
      Σ = 'sum of' (a mathematical symbol)

(b) DETERMINATION OF ROUGHNESS

Figure 4.6 Surface texture

(d) Optical instruments such as the autocollimator, the laser and the alignment telescope are also widely used for alignment setting, but are outside the scope of this book. They are considered within the main competencies.

Roundness and concentricity

Roundness   refers to the cross-section of a cylindrical or conical component being truly circular. Checking for roundness with a micrometer caliper may not show up out-of-roundness. Figure 4.9a shows a form of out-of-roundness (constant-diameter lobed figure) which will only show up when the component is supported in a V-block and rotated under a dial indicator, as shown in Figure 4.9b.

Concentricity implies a number of diameters with a common centre. There are various ways of testing for concentricity; two are shown in Figure 4.9c.

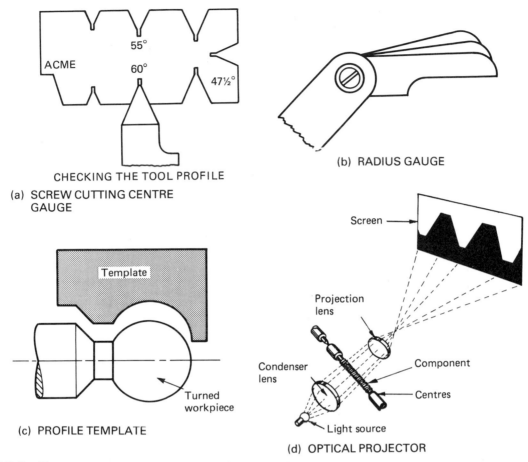

55°

ACME

60°

47½°

CHECKING THE TOOL PROFILE

(a) SCREW CUTTING CENTRE
    GAUGE

(b) RADIUS GAUGE

Template

Turned
workpiece

(c) PROFILE TEMPLATE

Screen

Projection
lens

Condenser
lens

Component

Centres

Light source

(d) OPTICAL PROJECTOR

Figure 4.7 Profiles

Accuracy of form

This is the combination of all the foregoing dimensional properties. At one time it was only necessary to specify the dimensional tolerances on a component drawing and to assume that the manufacturing process would produce the required accuracy of form. Nowadays the precision and the performance of engineering components is so much more exacting that this can no longer be assumed. Therefore geometrical as well as dimensional tolerancing has to be used on component drawings to ensure that components are the correct shape as well as the correct size. This leads to ease of assembly and uniform performance of engineering products.

4.3 Standards of measurement

Length

The rapid advances in technology made during the nineteenth century stimulated British and foreign governments into producing new and more easily read standards of length. However, these material standards varied in length over the years owing to molecular changes in the alloys from which they were made. Further, the American standard yard was very slightly longer than the British standard yard. An attempt was made to overcome these difficulties in 1960 by the adoption of an international yard based upon the metre. (The metre was, at that time, defined as the length of a

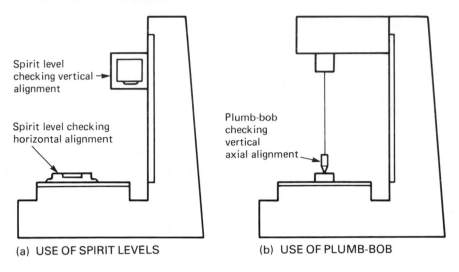

Spirit level checking vertical alignment

Spirit level checking horizontal alignment

(a) USE OF SPIRIT LEVELS

Plumb-bob checking vertical axial alignment

(b) USE OF PLUMB-BOB

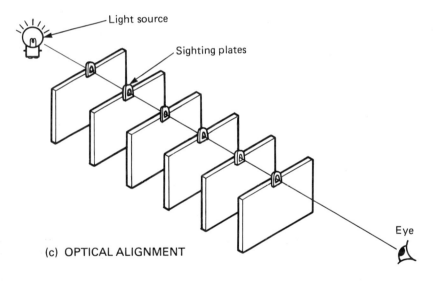

Light source

Sighting plates

Eye

(c) OPTICAL ALIGNMENT

Figure 4.8 Relative position

standard bar kept in Paris.) This international yard (0.9144 metres) became the British legal standard from 31 January 1964. Further, the metre itself was redefined at the Eleventh General Conference of Weights and Measures held in Paris in 1960. In place of the material standard the metre was redefined as equal to 1 650 763.73 vacuum wavelengths of the orange radiation of krypton isotope 86 gas when used in an electrical discharge tube.

However, technology increasingly demands more accurate standards, and the metre is now defined as the length of the path travelled by light in vacuum in 1/299 792 458 seconds. This can be realized in practice through the use of an iodine-stabilized helium-neon laser. The reproducibility is 3 parts in 10^{11}, which may be compared to measuring the earth's mean circumference to an accuracy of about 1 mm.

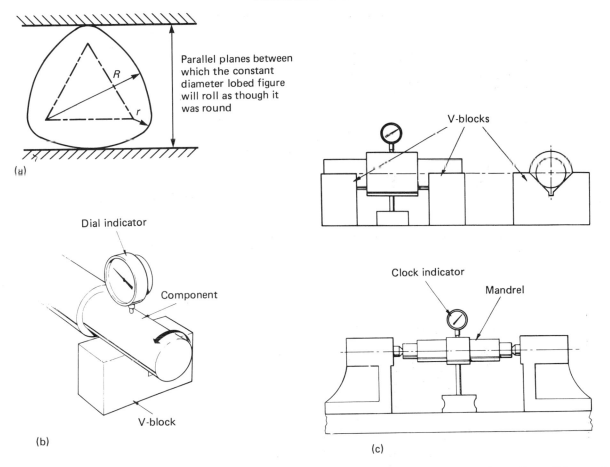

Figure 4.9 (a) Out-of-roundness which cannot be determined by micrometer or vernier calipers (b) use of V-block and DTI to determine out-of-roundness (c) testing for concentricity

Optical standards have major advantages over the older metal bars:

(a) They do not change length.
(b) Damaged or destroyed standards can be replaced without loss of accuracy.
(c) Identical copies can be kept in standards rooms and physical laboratories throughout the world.
(d) It is possible to make comparative measurements with the new standards of a much higher order of accuracy than was possible with the older material standards.

For practical measurements there is a hierarchy of working standards. For example, in most firms inspection-grade slip gauges would be used to check and measure work produced using workshop-grade slip gauges, micrometers, verniers, rules etc. The slip gauges would be calibrated against the laser by the National Physical Laboratory, and the firm would receive a calibration chart (a larger company might well have its own laser standard). The calibrated slip gauges could then, in turn, be used to check and calibrate micrometers, verniers and rules for day-to-day use in the workshop.

International standardization

As well as length, many other quantities had to be defined together with the specifications for equipment. In Britain this is the responsibility of the British Standards Institution (BSI). Internationally it is the responsibility of the International Organization for Standardization (ISO). Metrication in Britain has enabled the BSI to align its standards with the ISO. This adoption of international standards has been essential to ensure interchangeability of components and equipment made in different countries, and as a means of developing and expanding world trade.

To satisfy a typical product specification it is essential that:

(a) The quality specification and the units of measurement have been clearly and accurately defined
(b) The manufacturing process can operate within the accuracy specified
(c) The measuring equipment available during manufacture and subsequent inspection is also capable of operating within the accuracy specified
(d) The performance of the product and the performance testing routine is also specified.

Table 4.1 Base units

Quantity	Unit	Abbreviation
Length	metre	m
Mass	kilogram	kg
Time	second	s
Electric current	ampere	A
Temperature	kelvin	K
Luminous intensity	candela	cd
Amount of substance	mole	mol

Quantities and units (SI)

The base units in the Système International d'Unités (SI or metric units) are listed in Table 4.1, and some derived units are listed in Table 4.2. Sometimes these units are too large or too small for convenient use, and multiples or submultiples of the units are used instead. These are listed in Table 4.3, together with some examples.

Dimensional standards

The development of international standards of length for linear dimensions has already been introduced. At the same time standardization of dimensional control was also developing on an international basis both in Great Britain and in the USA. At first standardization was confined to screw threads and gauges, but eventually these early standards were superseded by a comprehensive standard for limits and fits. This, in turn, was eventually converted into an international metric standard of limits and fits (BS 4500).

Technical terms and symbols

Technical terms and symbols are also standardized to provide a universal language for industry in order to prevent errors through misinterpretation of instructions. Examples of standard technical terms and symbols are to be found in engineering drawings prepared to BS 308 (see Chapter 9).

Quality specifications for measuring tools and equipment

In addition to quality specifications for materials and components, there are British Standards for measuring tools and equipment such as rules,

Table 4.2 Some derived units

Quantity and symbol	Derivation		Unit	Unit symbol
Force F, weight G	mass × acceleration	(kg m/s^2)	newton	N
Stress p, pressure p	force/unit area	(N/m^2)	pascal	Pa
Work W, energy E	force × distance	(Nm)	joule	J
Power P	work/unit time	(J/s)	watt	W
Velocity v	distance/unit time	(m/s)	—	m/s
Acceleration a	velocity/unit time	(m/s^2)	—	m/s^2
Frequency f	cycles/unit time	(cycles/s)	hertz	Hz

Table 4.3 Multiples and submultiples

Multiple or submultiple	Prefix	Symbol
1 000 000 000 000 = 10^{12}	tera	T
1 000 000 000 = 10^9	giga	G
1 000 000 = 10^6	mega	M
1 000 = 10^3	kilo	k
1 = 10^0	—	—
0.001 = 10^{-3}	milli	m
0.000 001 = 10^{-6}	micro	μ
0.000 000 001 = 10^{-9}	nano	n
0.000 000 000 001 = 10^{-12}	pico	p

Examples
1 kilometre (km) = 1000 m
1 millimetre (mm) = 0.001 m
1 megahertz (MHz) = 1 000 000 Hz
1 microfarad (μF) = 0.000 001 F

micrometers, verniers, slip gauges, dial test indicators (DTIs) and surface plates. Such standards include not only the recommended tolerances for the manufacture of such devices, but also recommendations for their testing, care and use.

The accuracy of measuring equipment must be substantially greater than the dimensional feature being measured. A general rule is that the measuring instrument should be ten times more accurate than the dimension it is measuring.

Environmental standards in measurement

For precise measurement it is essential to house the equipment in a standards room which has a controlled environment. The internationally agreed temperature for measuring purposes is 20°C. The humidity should also be controlled to avoid corrosion, and the standards room should be free from dust as this can also cause errors and wear in the equipment. The need for a constant temperature is necessary because most materials expand as the temperature rises and contract as the temperature falls. This not only affects the linear dimensions of components, but can also cause twisting and distortion in large assemblies.

4.4 Advantages of standards

Nowadays we expect every M8 × 1.25 mm nut to fit every M8 × 1.25 mm bolt irrespective of the maker and country of origin (see Section 8.9 for screw thread systems). This is only possible by the international adoption of standard specifications. Standardized components can have other advantages. Batch sizes are increased, which reduces manufacturing costs. Assembly can be on a non-selective basis, and this also reduces costs, as does the reduced range of sizes and types of components it is necessary to keep in stock. Components made to an approved standard also have a uniform high quality and performance and are safe to use.

4.5 Measurement of length

Rules

These should be made from spring temper stainless or chrome-plated carbon steel. The finish should be matt and the graduations should be engine engraved for accuracy and clarity. The edges should be ground so that the rule can also be used as a straight edge. The datum end should also be ground, and should be used with care to prevent wear leading to inaccuracies. The rule should be thin to prevent sighting (parallax) errors in use. Figure 4.10 shows some typical techniques to ensure accuracy of use of rules.

Rules are made to high standards of precision and should be looked after carefully. They should be kept clean, and the edges should be kept free from burrs. They should never be used for scraping out swarf from the T-slots of machine tables, or as screwdrivers.

For longer distances, measuring tapes can be used.

Vernier calipers

Figure 4.11 shows a typical vernier caliper together with typical applications for internal and external measurements. Note that for internal measurements, the combined width of jaws has to be added to the scale reading. All vernier reading instruments rely upon very accurately engraved scales. The main scale is marked in standard increments of measurement similar to a rule (in the UK there are often two scales, one in metric and the other in imperial units). The vernier scale which slides

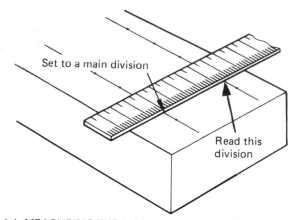

(a) MEASURING THE DISTANCE BETWEEN TWO
    SCRIBED LINES

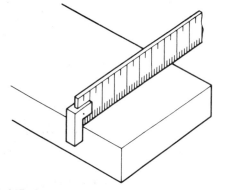

(b) MEASURING THE DISTANCE BETWEEN TWO
    FACES USING A HOOK RULE

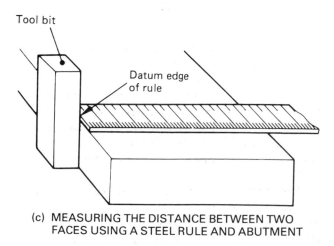

(c) MEASURING THE DISTANCE BETWEEN TWO
    FACES USING A STEEL RULE AND ABUTMENT

Figure 4.10 Use of the engineer's rule

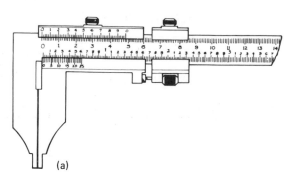

(a)

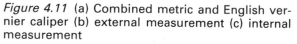

(b)

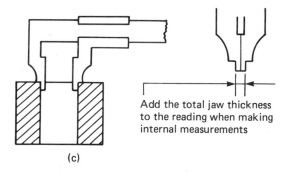

(c)

Figure 4.11 (a) Combined metric and English vernier caliper (b) external measurement (c) internal
measurement

along the main scale is marked with divisions
slightly smaller than the main scale divisions. It is
the difference between these divisions which
determines the accuracy to which the vernier can
be read.

The main scale of the metric vernier is marked
off in 1 mm increments and the vernier scale is
marked off in 0.98 mm divisions. The difference is
thus 0.02 mm, which is the reading accuracy of the
instrument. The reading is obtained by first noting
the distance that the zero on the vernier scale has
travelled along the main scale (23 mm in Figure

METRIC VERNIER

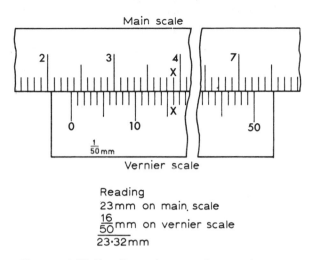

Reading
23mm on main. scale
$\frac{16}{50}$mm on vernier scale
23·32mm

Figure 4.12 Reading the metric vernier scale: 23 mm on main scale plus 16 × 0.02 mm on vernier scale gives total 23.32 mm

(a)

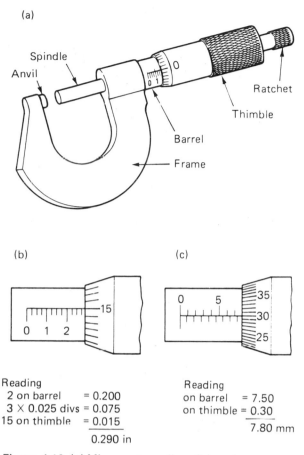

(b)

(c)

Reading
2 on barrel   = 0.200
3 × 0.025 divs = 0.075
15 on thimble   = 0.015
                           0.290 in

Reading
on barrel   = 7.50
on thimble = 0.30
                      7.80 mm

Figure 4.13 (a) Micrometer caliper (b) scales English scale reading: 0.275 on barrel plus 0.015 on thimble gives total of 0.290 inches
Metric scale reading: 7.50 on barrel plus 0.30 on thimble gives total 7.80 millimetres

4.12). Then add the number of 0.02 mm increments (in this case 16) counted along the vernier scale up to the point of coincidence X-X. This is where a line on the vernier scale coincides with a line on the main scale. The reading on the figure is therefore 23.32 mm.

Some verniers have scales different from the one described, and the scales should be checked before taking a reading. As with all measuring instruments, the vernier caliper must be treated carefully, and cleaned and returned to its case whenever it is not in use.

The vernier principle is applied to instruments other than the caliper, and the use of the vernier height gauge is introduced in Section 5.3.

Micrometer caliper

Figure 4.13a shows a typical micrometer caliper. This depends upon a precision screw and nut for its accuracy.

In a metric micrometer the thread employed has a pitch of 0.50 mm. The barrel is marked in 0.50 mm increments – that is, one increment per revolution of the screw thread and the thimble. The barrel markings are placed above and below the datum line alternately for ease of reading. The

thimble itself is marked with 50 equally spaced divisions, giving a reading accuracy of 0.01 mm (0.50 mm/50 = 0.01 mm).

For the English micrometer a thread having 40 turns per inch is used. The barrel is marked off in 1/40 inch (0.025 inch) increments. The thimble is marked with 25 equally spaced divisions, thus allowing the instrument to read to 1/1000 (0.001) inch.

Care must be taken to avoid straining the frame of the micrometer caliper by using excessive measuring pressure. Two or three clicks of the ratchet is sufficient. The micrometer should be kept clean and in its case when not in use, and the

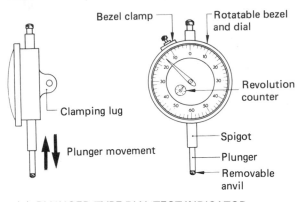

(a) PLUNGER-TYPE DIAL TEST INDICATOR

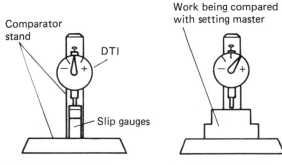

(i) DTI SET TO ZERO USING SLIP GAUGES AS A SETTING MASTER

(ii) ANY DEVIATION BETWEEN SIZE OF THE WORKPIECE AND THE SETTING MASTER IS INDICATED AS A + OR × READING ON THE DTI SCALE

(b) DTI AS A COMPARATOR

Figure 4.14 Use of dial test indicator (plunger type)

anvils should be kept slightly open. Before use the scales should be checked that they read zero when the anvils are closed under normal measuring pressure. If this is not so, then the barrel should be rotated in the frame with the C-spanner provided until a zero reading is obtained. Figure 4.13b shows an example of reading the scales of an English micrometer, and Figure 4.13c an example of reading a metric micrometer. Different types of micrometer are available for internal and depth measurements.

Dial test indicators (DTI)

There are two types of instrument in common use.

Plunger type

An example of this type is shown in Figure 4.14a. A gear train is used to magnify the displacement of the plunger, and the magnitude of its displacement is indicated by the pointer and scale. The small pointer counts the number of complete revolutions made by the main pointer. Various scales and magnifications are available.

Figure 4.14b shows a plunger DTI being used as a comparator. It indicates the deviation of the component from the size of the setting master, in this case a stack of slip gauges.

Lever type

An example of this type is shown in Figure 4.15a. A lever and scroll is used to magnify the displacement of the stylus. This type of instrument has only a limited range of movement compared with the plunger type. However, it is extremely popular for inspection and machine setting, partly because it is more compact, and partly because the scale and pointer are more conveniently positioned to be read in these applications.

An example of the use of a lever DTI is shown in Figure 4.15b.

4.6 Measurement of angles

Right angles

These are usually checked by comparison with a try-square. Figure 4.16 shows a typical try-square made to BS 939, and names its essential features. There are three grades of accuracy:

AA reference squares for use in standards rooms
A   precision squares for checking and inspection
B   general-purpose squares for workshop use.

There are various ways in which the right angle between two surfaces of a component may be compared with a try-square. Small components can be checked directly. The stock is slid down one surface of the component and, if the surfaces are at right angles, there should be no light gap between

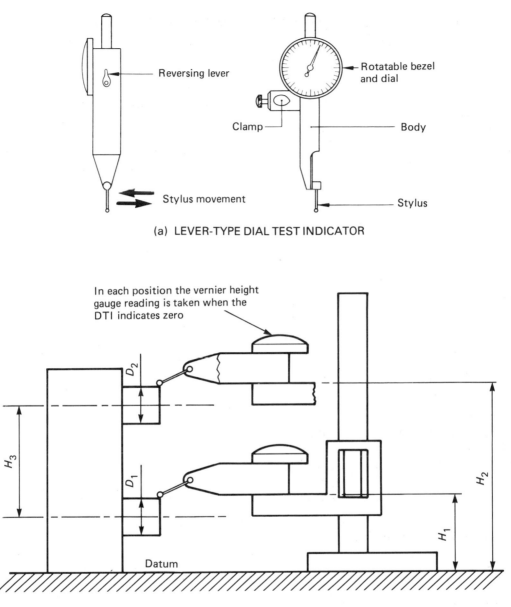

(a) LEVER-TYPE DIAL TEST INDICATOR

$H_3 = H_2 - H_1$ when $D_1 = D_2$

(b)   DTI AS A FIDUCIAL INDICATOR: $H_3 = H_2 - H_1$ WHEN $D_1 = D_2$: ZERO READINGS GUARANTEE CONSTANT MEASURING PRESSURE

Figure 4.15 Use of dial test indicator (lever type)

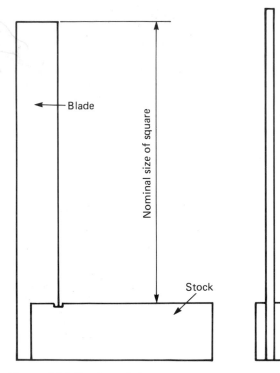

Figure 4.16 Engineer's try-square

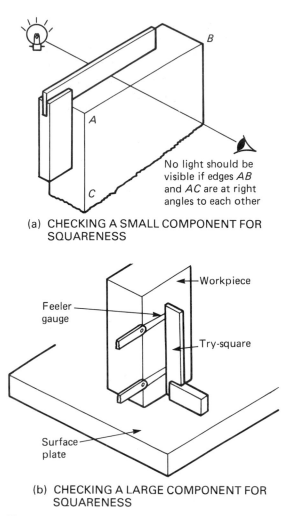

(a) CHECKING A SMALL COMPONENT FOR SQUARENESS

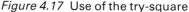

(b) CHECKING A LARGE COMPONENT FOR SQUARENESS

Figure 4.17 Use of the try-square

the blade and the other surface (Figure 4.17a). Larger components can be checked as shown in Figure 4.17b. If the surfaces are at right angles the gap between the blade and the component should be constant when it is checked with feeler gauges as shown.

In addition to try-squares, cylinder squares and prismatic squares may also be used for checking large work. The form of these squares ensures a line contact between them and the work. They are used in a similar manner to the example in Figure 4.17b, or they may be used in conjunction with a squareness comparator as shown in Figure 4.18. The DTI is set to zero against a master gauge such as the cylinder square shown in the figure. Any out-of-squareness in the component is shown up as a plus or minus reading on the DTI.

Angles other than right angles

Angles other than a right angle can be measured using a bevel protractor. This is shown in Figure 4.19a. The reading accuracy of this instrument is limited to 30 minutes of arc (0.5 degrees). For more accurate measurement the vernier protractor is used. The twelve-division vernier scale shown in Figure 4.19b has a reading accuracy of 5 minutes of arc.

4.7 Dimensional deviation

Dimensional deviation has to be allowed because it is not possible to manufacture a component to an exact size, and it is not possible to measure an exact size. Therefore the designer sets the limits of

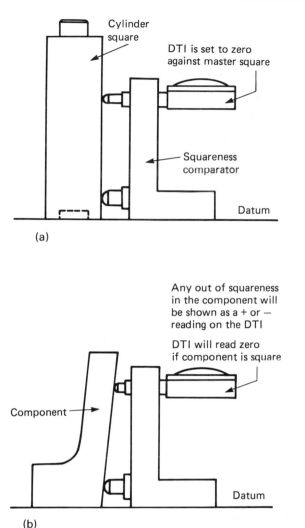

Figure 4.18 (a) Setting squareness comparator (b) Checking component for squareness

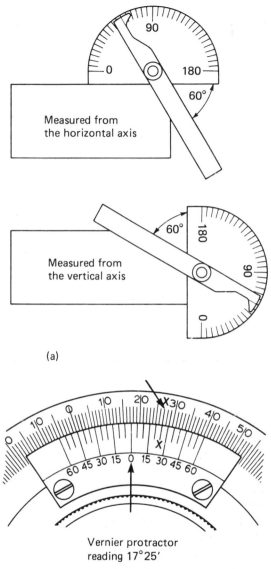

Figure 4.19 Measurement of angles (a) bevel protractor (b) vernier protractor

size between which the component may be manufactured, yet still function satisfactorily.

Nominal size   This is the dimension by which the feature is identified for convenience.

Limits   The limits of size are the high and low values of size between which it is permitted to make the component.

Tolerance   This is the arithmetic difference between the upper limit of size and the lower limit of size. Tolerances may be *bilateral*, in which case the tolerance band is partly above and partly below the nominal size, or *unilateral*, in which case the tolerance band is entirely above or entirely below the nominal size. Table 4.4 shows standard tolerance grades.

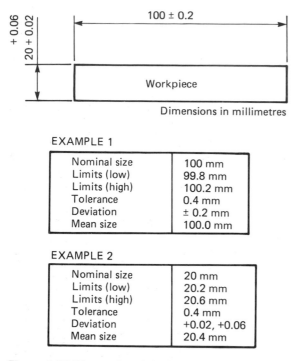

EXAMPLE 1

Nominal size	100 mm
Limits (low)	99.8 mm
Limits (high)	100.2 mm
Tolerance	0.4 mm
Deviation	± 0.2 mm
Mean size	100.0 mm

EXAMPLE 2

Nominal size	20 mm
Limits (low)	20.2 mm
Limits (high)	20.6 mm
Tolerance	0.4 mm
Deviation	+0.02, +0.06
Mean size	20.4 mm

Figure 4.20 Dimensional deviation

Deviation This is the difference between the nominal size and either limit. The deviation may be *symmetrical*, in which case the deviation on either side of the nominal size is the same, or *asymmetrical*, in which case the deviation may be greater on one side of the nominal size than on the other.

Mean size This dimension lies halfway between the upper and lower limits of size and should not be confused with the nominal size. They are only the same when bilateral tolerance and symmetrical deviation are used at the same time. When setting up a machine it is normal to adjust it so that the components are produced at the mean size so that any normal variation in size during manufacture will maintain the component within limits.

Figure 4.20 shows how these terms are applied in dimensioning a component.

4.8 Accuracy

The narrower the tolerance band, the greater will be the accuracy of the component, and the more

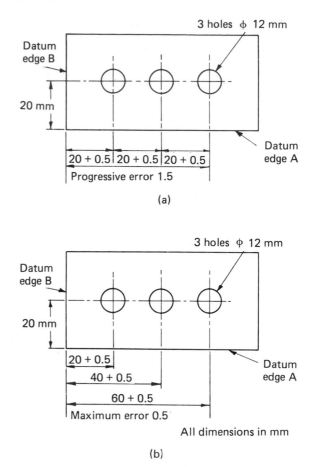

Figure 4.21 Effect of dimensioning on accuracy: (a) chain dimensioning (b) dimensioning from a datum edge

difficult and costly it will be to manufacture the component within the limits specified. Therefore the designer never specifies an accuracy greater than that which is necessary to ensure correct function of the component. Table 4.5 relates various manufacturing processes to the accuracy which can be expected from them.

The method of dimensioning can also affect the accuracy of the component. Figure 4.21a shows how chain dimensioning (incremental dimensioning) can lead to a *cumulative error* which will be greater than the designer intends. Figure 4.21b shows how dimensioning from a datum (absolute dimensioning) can overcome this build-up of error.

Table 4.4 Standard tolerances: tolerance unit 0.001 mm

Nominal sizes		Tolerance grades																	
Over (mm)	Up to and including (mm)	IT01	IT0	IT1	IT2	IT3	IT4	IT5	IT6†	IT7	IT8	IT9	IT10	IT11	IT12	IT13	IT14*	IT15*	IT16*
—	3	0.3	0.5	0.8	1.2	2	3	4	6	10	14	25	40	60	100	140	250	400	600
3	6	0.4	0.6	1	1.5	2.5	4	5	8	12	18	30	48	75	120	180	300	480	750
6	10	0.4	0.6	1	1.5	2.5	4	6	9	15	22	36	58	90	150	220	360	580	900
10	18	0.5	0.8	1.2	2	3	5	8	11	18	27	43	70	110	180	270	430	700	1100
18	30	0.6	1	1.5	2.5	4	6	9	13	21	33	52	84	130	210	330	520	840	1300
30	50	0.6	1	1.5	2.5	4	7	11	16	25	39	62	100	160	250	390	620	1000	1600
50	80	0.8	1.2	2	3	5	8	13	19	30	46	74	120	190	300	460	740	1200	1900
80	120	1	1.5	2.5	4	6	10	15	22	35	54	87	140	220	350	540	870	1400	2200
120	180	1.2	2	3.5	5	8	12	18	25	40	63	100	160	250	400	630	1000	1600	2500
180	250	2	3	4.5	7	10	14	20	29	46	72	115	185	290	460	720	1150	1850	2900
250	315	2.5	4	6	8	12	16	23	32	52	81	130	210	320	520	810	1300	2100	3200
315	400	3	5	7	9	13	18	25	36	57	89	140	230	360	570	890	1400	2300	3600
400	500	4	6	8	10	15	20	27	40	63	97	155	250	400	630	970	1550	2500	4000
500	630	—	—	—	—	—	—	—	44	70	110	175	280	440	700	1100	1750	2800	4400
630	800	—	—	—	—	—	—	—	50	80	125	200	320	500	800	1250	2000	3200	5000
800	1000	—	—	—	—	—	—	—	56	90	140	230	360	560	900	1400	2300	3600	5600
1000	1250	—	—	—	—	—	—	—	66	105	165	260	420	660	1050	1650	2600	4200	6600
1250	1600	—	—	—	—	—	—	—	78	125	195	310	500	780	1250	1950	3100	5000	7800
1600	2000	—	—	—	—	—	—	—	92	150	230	370	600	920	1500	2300	3700	6000	9200
2000	2500	—	—	—	—	—	—	—	110	175	280	440	700	1100	1750	2800	4400	7000	11000
2500	3150	—	—	—	—	—	—	—	135	210	330	540	860	1350	2100	3300	5400	8600	13500

* Not applicable to sizes below 1 mm
† Not recommended for fits in sizes above 500 mm

The ratio of tolerance to the nominal dimension is also important. A tolerance of 0.5 mm on a dimension of 2500 mm has a greater accuracy than a tolerance of 0.1 mm on a nominal dimension of 25 mm. Thus the tolerance must be carefully matched to the nominal size. If the designer works to BS 4500: Part 1: 'General tolerances and deviations', all these factors are taken into account for a range of dimensions from zero 3150 mm.

4.9 Factors affecting accuracy

The main factors affecting accuracy when measuring are as follows.

Temperature

All metals expand when heated and contract when cooled. Therefore a metal component which became very hot whilst being machined will reduce in size on cooling down to room temperature and

Table 4.5 Tolerance grades and manufacturing processes

IT number	Class of work
16	Sand casting and flame cutting
15	Stamping
14	Die casting and plastic moulding
13	Presswork and extrusion
12	Light presswork and tube drawing
11	Drilling, rough turning, boring
10	Milling, slotting, planing, rolling
9	Low-grade capstan and automatic lathework
8	Centre lathe, capstan and automatic lathework

may become undersize. Casting patterns have to be made oversize to allow for shrinkage, so that when the molten metal solidifies and cools the casting will not be undersize. Contraction rules are available which have a normal scale on one side and an expanded scale on the other to make allowance for this shrinkage. Even more important than direct expansion or contraction is the distortion which can occur when an assembly is made from a variety of materials all of which have different coefficients of linear expansion.

Accuracy of equipment

It has already been stated that it is impossible to manufacture components exactly to size or to measure them exactly. This equally applies to measuring instruments, so they also have to be manufactured within prescribed tolerances. As a general rule, measuring instruments should be ten times more accurate than the dimension being measured. Measuring standards such as slip gauges are frequently calibrated against even more accurate standards, and the deviations in size are charted. This enables any cumulative error to be calculated and allowed for when wringing slip gauges together into a stack.

Reading errors

Two different reading errors can occur:

Misreading the instrument scales. Verniers are particularly difficult to read, and the use of a magnifying glass is advised.
Sighting errors (parallax) when using rules and similar scales. The eye should be immediately over the point of measurement. The use of a solid abutment as a datum is shown in Figure 4.10; this avoids the necessity of having to sight two points at the same time.

Type of equipment

It is possible to measure linear dimensions with a variety of instruments. However, the accuracy of measurement is always of a lower order than the reading accuracy and will depend upon the skill of the user.

Line measurement

Line measurement takes place, as its name suggests, when the distance between two lines or edges is measured with an instrument such as a steel rule. The following should be noted:

(a) The distance between two scribed lines is particularly difficult to measure. The problem of sighting errors has already been discussed above, and datums cannot be used in this instance. A pair of dividers can be used to transfer the dimension. Since one leg can be positively located in one main division of the rule, only one reading has to be sighted.
(b) The distance between two edges is also difficult to measure using line measurement techniques. The use of a solid abutment as an aid to using a rule is shown in Figure 4.10, and the use of calipers to transfer a dimension is shown in Figure 4.22

End measurement

Since the engineer is usually concerned with the measurement of distances between surfaces, end measurement is to be preferred. End measurement is the name given to measuring techniques which use instruments such as the vernier caliper and the micrometer caliper, where the end surfaces of the component are in contact with the jaws of the measuring instrument and no sighting is required.

Application of force

Excessive force when closing the measuring instrument on the workpiece can result in the workpiece being distorted and/or the measuring instrument being distorted. Force should be applied sufficient just to feel the workpiece between the jaws of the instrument.

Some instruments are fitted with devices for applying the correct force. For example, with the ratchet of the micrometer caliper, three clicks are generally considered to indicate sufficient force. The bench micrometer has a measuring force indicator (fiducial indicator) in place of the fixed anvil, as shown in Figure 4.23. The measuring pressure when using a DTI is limited to the strength of the plunger return spring.

The contact pressure can also vary despite the use of a constant measuring force. Pressure is defined as force per unit area, and a reduction in area for a given force can result in a large increase in pressure. In theory, the point contact associated

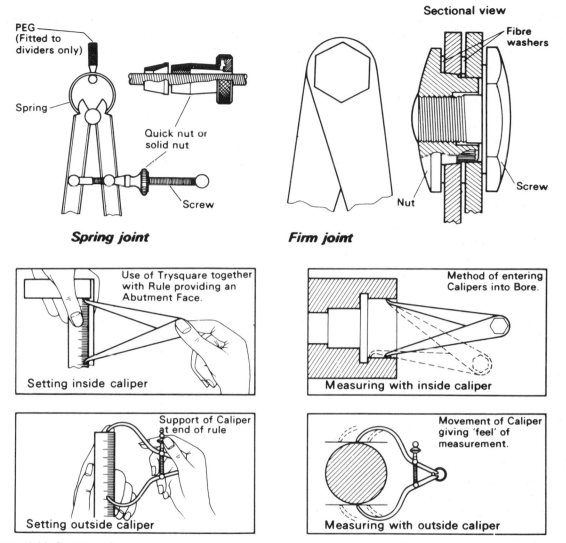

Figure 4.22 Construction and use of calipers (courtesy Moore & Wright Ltd)

with spherical components results in an infinitely large measuring pressure. In practice, there is bound to be some slight indentation of the surface being measured and some slight flattening of the measuring instrument contact faces until the build-up in contact area reduces the measuring pressure to an acceptable value.

Use of measuring equipment

(a) The measurement must be made at right angles to the surfaces of the component.

(b) The component must be supported so that it does not collapse under the measuring pressure or under its own weight.

(c) The workpiece must be cleaned before being measured, and coated with oil or a corrosion inhibitor after inspection.

(d) Measuring instruments must be handled with care so that they are not damaged or strained. They must be kept in their cases when not in use and kept clean and lightly oiled on the bright surfaces. They should be regularly checked to ensure that they have not lost their initial accuracy.

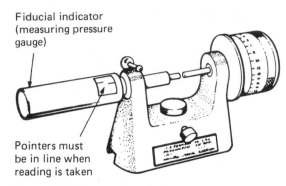

Fiducial indicator
(measuring pressure
gauge)

Pointers must
be in line when
reading is taken

Figure 4.23 Bench micrometer

4.10 Terminology of measurement

Indicated size   This is the size indicated by the scales of the measuring instrument when a measurement is being taken. It makes no allowance for incorrect use of the instrument, such as the application of excessive force.

Reading   This is the operator's interpretation of the indicated size. Vernier scales are easily misread in bad light. The digital readouts of electronic measuring instruments are easier to read and errors are less likely to occur.

Reading value   This is the smallest increment of size which can be read directly from the scales of the instrument. For a metric micrometer it is 0.01 mm.

Measuring range   This is the range of size between the largest and smallest dimension which can be read by a measuring instrument. For example, the measuring range of a 25 mm to 50 mm micrometer is $50 - 25 = 25$ mm.

Measuring accuracy   This is the accuracy of measurement expected from a measuring instrument after taking into account all the factors considered in Section 4.9. It can never be better than the indicated size. It is also called the tool accuracy, and is defined as the maximum allowable deviation relative to the indicated measurement.

4.11 Miscellaneous equipment

Surface plates and tables

These provide a datum surface for use during measuring and marking out operations (see also Chapter 5). A typical cast iron surface plate has heavy ribbing incorporated in its underside to give great rigidity for a given mass of metal. Surface tables are of similar structure but very much larger, and they are free standing.

Cast iron is used for the following reasons:

(a) It is self-lubricating, and the equipment slides on its working surface with a pleasant feel.
(b) It is easily cast to provide the complex shape of the stiffening ribs.
(c) It is a stable and rigid metal and relatively inexpensive.
(d) It is easily machined and scraped to an accurate plane surface.

Granite is also used for large surface tables. This material is also very dense and stable and has the added advantage that, if scratched, it does not throw up a burr like cast iron.

Examples of the use of surface plates are shown in Chapter 5.

V-blocks

These are used to support cylindrical components: an example is shown in Figure 4.24. They may also be used to support rectangular components at 45° to the datum surface. V-blocks are manufactured in pairs so that long components can be supported parallel to the datum surface, and for this reason they must always be bought and kept as a pair.

Spirit levels

A precision block level is shown in Figure 4.8a. The sensitivity of a level depends upon the radius of the vial: the greater this radius, the greater the sensitivity. The sensitivity of spirit levels may be expressed in two ways:

(a) A 10 second level means that tilting the level through an angle of 10 seconds to the horizontal will move the bubble one division.
(b) A level with a sensitivity of 0.05 mm/m means that if the level were placed on a one metre long straight edge, and that if one end of that straight edge was raised by 0.05 mm, then the bubble would move one division.

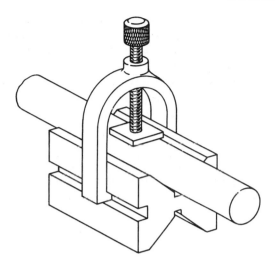

Figure 4.24 V-block and clamp

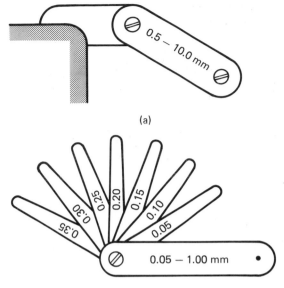

(a)

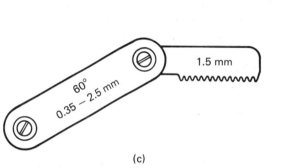

(b)

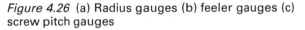

(c)

Figure 4.26 (a) Radius gauges (b) feeler gauges (c) screw pitch gauges

Levels need to be carefully used and cared for to prevent damage. They must never be dropped, and should be kept in their cases when not in use. The ground surfaces must be cleaned before and after use, and oiled before the level is returned to its case.

When a level is used, two readings should always be taken; the level is turned through 180° between readings. The true reading is the mean of the two readings. If there is an error of more than one division, the vial should be adjusted until there is virtually no error when the level is reversed.

Straight edges

These may be rectangular or bevel edged. They are used for testing for straightness and flatness.

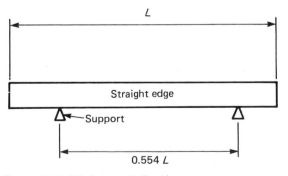

Figure 4.25 Minimum deflection

They may also be used as a guide for a scriber when marking out straight lines. Long straight edges tend to bend under their own weight and need to be supported at the points of minimum deflection, as shown in Figure 4.25. Note these are *not* the same as Airey point (0.577*L*) used to support material standards of length.

Gauges

Radius gauges These are supplied in sets providing a range of internal and external radii. A typical application is shown in Figure 4.26a.

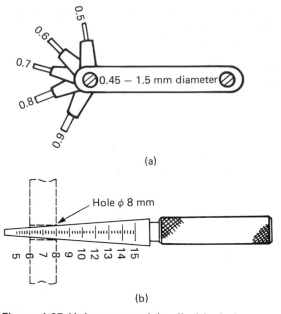

(a)

Hole φ 8 mm

(b)

Figure 4.27 Hole gauges: (a) cylindrical pin gauges (b) taper pin gauge for hole sizes 5–15 mm diameter

(a) Measuring equipment should be between twice and ten times as accurate as the dimension being measured. The latter figure should be aimed for.

(b) Accuracy is improved if the setting master used for comparative measurement is as close as possible to the dimension being measured.

(c) Wherever possible, measurements should be taken at the standard temperature of 20°C.

(d) Measuring equipment must be carefully used and looked after. It must be cleaned before and after use. It must not be mixed with other bench tools and cutting tools. It should be kept in the cases provided when not in use, and bright surfaces should be protected by a film of oil or petroleum jelly. Measuring equipment should be inspected for errors or damage before use, and there should be a regular programme of maintenance and re-calibration.

(e) The maker's recommendations for maintaining equipment in good condition must be observed.

Feeler gauges These are used to measure clearances between components. Figure 4.26b shows a typical set of feeler gauges. A typical range of thickness is from 0.05 mm to 1.00 mm in increments of 0.05 mm. Other ranges of thickness and increment are available. A typical application is shown in Figure 4.17b.

Screw pitch gauges These are used for checking the pitch of a screw thread. They are available with 55° and 60° included thread angles, and in metric and inch pitches. A typical set is shown in Figure 4.26c.

Hole gauges Sets of cylindrical pin gauges are available (Figure 4.27a) for checking the diameter of small holes. In a typical set the pins range from 0.45 mm to 1.5 mm in steps of 0.05 mm. For larger holes, taper pin gauges are available (Figure 4.27b).

4.12 General rules for accurate measurement

The general rules for accurate measurement may be summarized as follows:

Exercises

For each exercise, select *one* of the four alternatives.

1 A 'go' and 'not go' caliper gauge is used to
   (a) determine the exact size of a component
   (b) check whether a dimension lies within the design limits
   (c) measure a thickness
   (d) measure an angle.

2 If two surfaces are at 90° to each other they are said to be
   (a) mutually perpendicular
   (b) mutually parallel
   (c) at an acute angle to each other
   (d) at an obtuse angle to each other.

3 Flatness is the extent to which the surface of a component
   (a) is smooth
   (b) deviates from a true plane
   (c) is free from machining marks
   (d) is horizontal.

4 Concentricity implies
  (a) lack of ovality
  (b) roundness
  (c) lack of taper in turned components
  (d) a number of diameters with a common centre.

5 The ISO standard of length is the
  (a) metre
  (b) imperial standard yard
  (c) international standard yard
  (d) millimetre.

6 The ISO recommended temperature for standards rooms is
  (a) 70°F
  (b) 20°C
  (c) 65°F
  (d) 18°C.

7 The screw thread specification M20 × 2.5 means
  (a) 2.5 mm diameter and 20 threads per metre
  (b) 20 mm diameter and 2.5 threads per millimetre
  (c) 20 mm diameter and 2.5 millimetre pitch
  (d) 2.5 mm diameter and 20 micrometer pitch.

8 The main purpose of the parallel strip shown in Figure 4.28 is to
  (a) prevent wear of the end of the rule
  (b) compensate for wear in the end of the rule
  (c) increase accuracy of measurement by reducing sighting errors
  (d) ensure the rule is at right angles to the component.

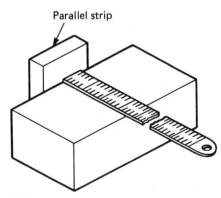

Figure 4.28

9 The reading of the micrometer caliper scales shown in Figure 4.29 is
  (a) 6.72 mm
  (b) 8.22 mm
  (c) 6.22 mm
  (d) 8.72 mm

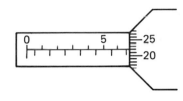

Figure 4.29

10 The reading of the vernier caliper scales shown in Figure 4.30 is
  (a) 47.00 mm
  (b) 33.28 mm
  (c) 33.14 mm
  (d) 47.14 mm.

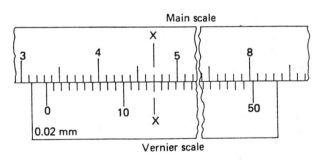

Figure 4.30

11 When making internal measurements with a vernier caliper, the jaw thickness is
  (a) ignored
  (b) subtracted from the reading
  (c) read on a separate scale
  (d) added to the reading.

12 The magnification of the stylus movement in the DTI shown in Figure 4.31 is achieved by
  (a) a gear train
  (b) a lever and scroll system
  (c) a rack and pinion
  (d) electronic amplification.

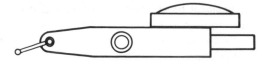

Figure 4.31

13 Figure 4.32 shows a DTI being used to determine the thickness of a component. The DTI
   (a) measures the thickness direct
   (b) compares the thickness of the component with a setting standard (slip gauge) and indicates the difference
   (c) measures the tolerance on the thickness direct
   (d) compares the thickness of the component with a setting master (slip gauge) and indicates the actual size.

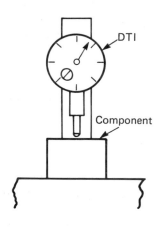

Figure 4.32

14 The most suitable instrument for measuring an angle of 15° 10′ between two surfaces is a
   (a) try-square
   (b) bevel protractor
   (c) vernier protractor
   (d) micrometer protractor.
15 A squareness comparator measures deviation from the perpendicular in
   (a) degrees
   (b) degrees and decimal fractions of a degree
   (c) millimetres
   (d) decimal fractions of a millimetre.

16 The nominal size of the hole shown in Figure 4.33 is
   (a) 25.15 mm
   (b) 25.05 mm
   (c) 25 mm
   (d) 0.10 mm.

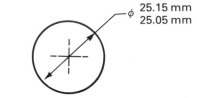

Figure 4.33

17 The tolerance on the hole shown in Figure 4.33 is
   (a) 25.15 mm
   (b) 25.05 mm
   (c) 25 mm
   (d) 0.10 mm.
18 The tolerance on the hole shown in Figure 4.33 is
   (a) unilateral
   (b) bilateral
   (c) multilateral
   (d) polylateral.
19 The lower limit of size shown in Figure 4.33 is
   (a) 25.15 mm
   (b) 25.05 mm
   (c) 25 mm
   (d) 0.10 mm.
20 The mean size for the dimension shown in Figure 4.33 is
   (a) 25.15 mm
   (b) 25.10 mm
   (c) 25.05 mm
   (d) 25.00 mm.
21 The narrower the tolerance band of a dimension,
   (a) the greater will be the accuracy and the less will be the cost
   (b) the less will be the accuracy and the less will be the cost
   (c) the greater will be the accuracy and the greater will be the cost
   (d) the less will be the accuracy and the greater will be the cost.

22 Measuring and gauging equipment should have
   (a) a greater accuracy than the feature being measured
   (b) the same accuracy as the feature being measured
   (c) a lower accuracy than the feature being measured
   (d) an accuracy which is unrelated to the feature being measured.

23 Chain dimensioning (incremental dimensioning) can
   (a) lead to accumulative errors
   (b) lead to reduced errors
   (c) lead to greater accuracy than when dimensioning from a datum
   (d) have no effect on the accuracy.

24 Excessive measuring pressure is
   (a) required to ensure firm contact with the workpiece
   (b) likely to cause errors due to distortion in the workpiece and instrument
   (c) likely to have no effect on the accuracy of measurement
   (d) essential if there is oil on the surface of the workpiece.

25 The reading value for a measuring instrument is
   (a) the operator's interpretation of the indicated size
   (b) the smallest increment which can be read directly from the scales
   (c) the measuring accuracy of the instrument
   (d) never better than the indicated size.

26 The measuring accuracy of an instrument is
   (a) always better than the indicated size
   (b) unrelated to the indicated size
   (c) always the same as the indicated size
   (d) never better than the indicated size.

27 A suitable material for surface plates and tables needs to be rigid, to be easily cast to shape, to be easily finished to a plane surface, to have a relatively low cost and to have anti-friction properties. Such a material is
   (a) mild steel
   (b) phosphor bronze
   (c) cast iron
   (d) grano-concrete.

28 The gauge shown in Figure 4.34 is used to check
   (a) the pitch of screw threads
   (b) the major diameter of screw threads
   (c) the thread angle of screw threads
   (d) the minor diameter of screw threads.

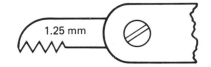

1.25 mm

Figure 4.34

29 V-blocks are mainly used to support
   (a) triangular work
   (b) rectangular work
   (c) conical work
   (d) cylindrical work.

30 The sensitivity of spirit levels increases when
   (a) the radius of curvature of the vial is made larger
   (b) the radius of curvature of the vial is made smaller
   (c) the length of the vial is made larger
   (d) the length of the vial is made smaller.

5

Marking out

5.1 Purposes, advantages and disadvantages of marking out

The purposes and advantages of marking out are as follows:

(a) To provide guide lines to work to, and to provide the only control of the size and shape of the component and the position and size of any features such as holes required in the component. This is satisfactory only where limited accuracy is required.
(b) As a guide to the machinist when setting up and cutting. In this instance the final dimensional control would come from the use of precision measuring instruments in conjunction with the micrometer dials on the machine controls.
(c) To ensure that adequate machining allowance has been left upon casting and forging. In addition, to ensure that such features as flanges, webs and cores are correctly positioned, and that cored holes are positioned centrally in their bosses and will clean up during machining.

The disadvantages of marking out are as follows:

(a) Working to a scribed line is only suitable for work of limited accuracy.
(b) Scribed lines cut into the surface of a component and deface the surface. Where the surface finish is important, the surface must be ground to remove the scribing marks.

(c) Drawn lines are too thick and indistinct to be worked to if any degree of precision is required.
(d) Centre punched holes may not provide sufficient accuracy for controlling the position of a hole axis.

5.2 Methods of marking out

It has been stated above that the purpose of marking out is to provide guide lines on the material being cut. Such lines may be drawn, scribed or centre marked.

Drawn lines

These are produced by a pencil; the point leaves a mark on the surface of the material, but does not cut into the surface. Such a mark is not permanent and is easily removed. Further, since the marking point is softer than the material being marked out, the point quickly wears and it is impossible to maintain a fine precise line. Drawn lines are only used when marking out tin plate or galvanized sheet steel. This is so that the line will not cut through the protective coating of tin or zinc and allow corrosion to take place.

Scribed lines

These are the clean, fine lines required for accurate working. A scriber (Figure 5.1a) is used to produce the line. It has a hard sharp point and cuts into the surface of the material, leaving a mark

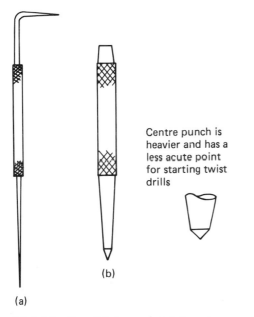

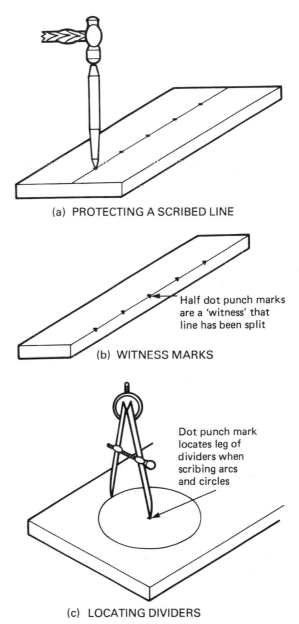

(a) PROTECTING A SCRIBED LINE

(b) WITNESS MARKS

(c) LOCATING DIVIDERS

Centre punch is heavier and has a less acute point for starting twist drills

Half dot punch marks are a 'witness' that line has been split

Dot punch mark locates leg of dividers when scribing arcs and circles

Figure 5.1 (a) Scriber (b) dot punch (c) centre punch

Figure 5.2 Use of the dot punch

which cannot be removed except by reforming the surface of the material (e.g. by surface grinding).

To ensure that the scribed line shows up clearly against the background, the surface of the material being marked out is usually coated thinly in a contrasting colour. For example, castings and forgings which have a surface covered in dark scale (oxide) are usually whitewashed before marking out. Bright surfaces may be given a thin coating of cellulose lacquer, and bright carbon steel surfaces can be treated with acidulated copper sulphate solution to 'copper plate' the surface being marked out. Care must be taken in the latter case since this solution is corrosive and will attack any instruments which may come into contact with it.

Centre marks

These may be made with either a centre punch or a dot punch as shown in Figure 5.1b. The centre punch (Figure 5.1c) is usually heavier and has a less acute point; it is used for making a mark for guiding the chisel point of a twist drill. The dot punch is lighter and has a finer point (60°), and is used for two purposes when marking out:

(a) A scribed line can be protected by a series of

centre marks made along the line, as shown in Figure 5.2a. If the line is accidentally removed in any way, it can be replaced by joining up the centre marks.

(b) Further, if the line is a guide for cutting, proof that the machinist has split the line is provided by the half-marks which remain, as shown in

Figure 5.2b. These remaining marks are often referred to as witness marks.

(c) Finally, dot punch marks are used to prevent the centre point of dividers from slipping when scribing circles and arcs of circles, as shown in Figure 5.2c.

When a·centre punch is driven into the work, distortion may occur. This can be a burr rising around the mark, swelling of the edges of the component, or buckling of thin components.

5.3 Marking-out equipment

The basic requirements for marking out are a scriber to produce a straight line, a rule to measure

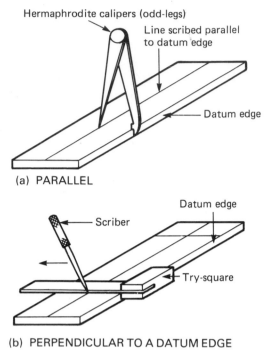

(a) PARALLEL

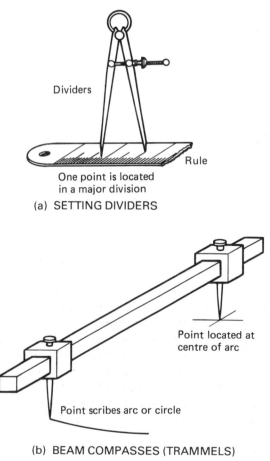

Dividers

One point is located
in a major division

(a) SETTING DIVIDERS

Point located at
centre of arc

Point scribes arc or circle

(b) BEAM COMPASSES (TRAMMELS)

Figure 5.3 Using dividers

(b) PERPENDICULAR TO A DATUM EDGE

Figure 5.4 Scribing lines

distances and to act as a straight edge to guide the scriber, and dividers for scribing circles and arcs of circles. The scriber and rule have already been described. Figure 5.3a shows how dividers are set; as can be seen, one reason why an engineer's rule should be engraved is so that the points of dividers can be clicked into the engraved marks to ensure their accurate setting. For large circles and arcs, beam compasses (trammels) are used, as shown in Figure 5.3b. The divider point is sharp enough to make its own centre point mark in the softer non-ferrous metals and most plastic materials. For harder materials it is often necessary to make a light centre mark with a dot punch to prevent the dividers slipping.

For most marking-out operations the scribed lines are either parallel to a datum edge or perpendicular to a datum edge. Figure 5.4a shows how odd-leg or hermaphrodite calipers are used to scribe a line parallel to a datum edge. Figure 5.4b shows how a try-square is used to guide the scriber point at right angles (perpendicular) to a datum

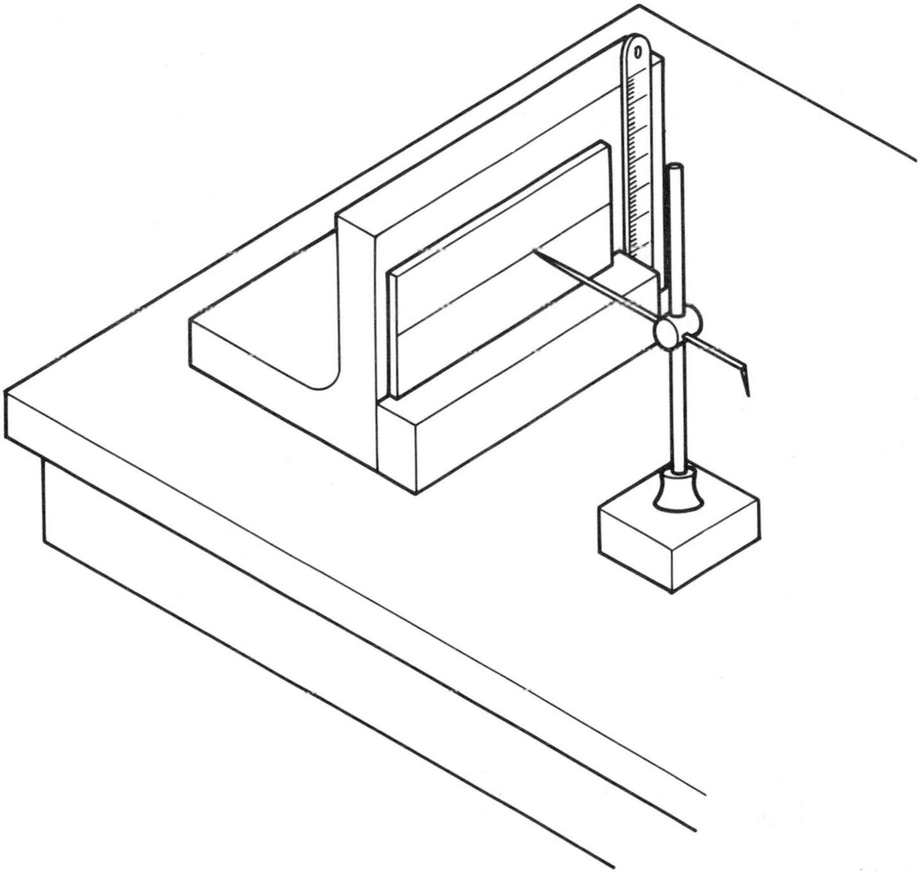

Figure 5.5 Marking out on the surface table

edge. For lines at some other angle to the datum edge, a bevel protractor may be substituted for the try-square. The set-up shown in Figure 5.5 may be used where greater precision is required and larger components are to be marked out.

In addition to the solid parallels shown in Figure 5.5, hollow cast iron parallel blocks are used where larger sizes are required. Such blocks are accurately machined and ground. Adjustable angle plates are also available where the component must be set at an angle other than a right angle. Screw-jacks and wedges are also used to support large and heavy work.

Flat plate components

Figure 5.5 shows a flat plate component set up for

marking out. It can be seen that there are three essential groups of equipment, as follows.

Those which are supporting the component
In this example a surface table, an angle plate and a precision-ground parallel packing strip are being used to support the component so that its datum edge is parallel to the surface table. The angle plate ensures that the surface being marked out is perpendicular to the surface table. In this instance the surface table is providing a common datum or plane from which all dimensions are taken.

Those from which dimensions are transferred
In this example a rule is being used as a basis of measurement for setting the scriber at the required

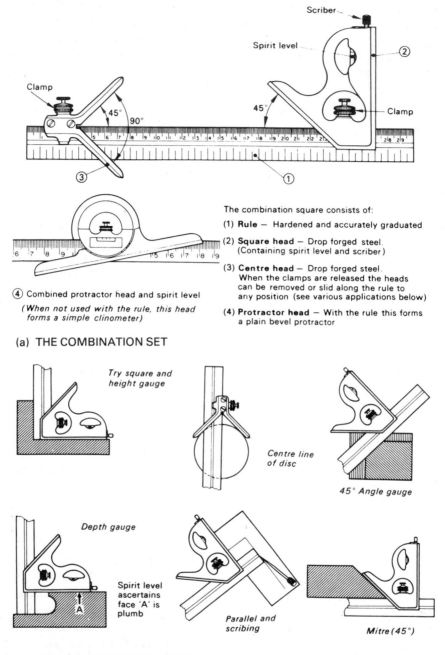

The combination square consists of:

(1) **Rule** – Hardened and accurately graduated

(2) **Square head** – Drop forged steel. (Containing spirit level and scriber)

(3) **Centre head** – Drop forged steel. When the clamps are released the heads can be removed or slid along the rule to any position (see various applications below)

(4) **Protractor head** – With the rule this forms a plain bevel protractor

(4) Combined protractor head and spirit level

(When not used with the rule, this head forms a simple clinometer)

(a) THE COMBINATION SET

Try square and height gauge

Centre line of disc

45° Angle gauge

Depth gauge

Spirit level ascertains face 'A' is plumb

Parallel and scribing

Mitre (45°)

(b) USES OF THE COMBINATION SET   (Courtesy of Moore and Wright Ltd)

Figure 5.6 A complete combination set (Courtesy of Moore and Wright Ltd)

height above the datum surface. The datum end of the rule must be in contact with the surface table. A combination square and rule could also be used, and a complete combination set is shown in Figure 5.6. It consists of a strong steel rule on to which can be fitted the following attachments:

Mitre square   This can be used as a try-square or a mitre square for marking out angles at 90° and 45° respectively.

Centre square   This is used for finding and marking the position of the axial centre of cylindrical components and circular blanks.

Bevel protractor   This is a plain protractor of limited accuracy, but is useful for scribing lines at angles other than a right angle. It is also fitted with a bubble level for setting work on an adjustable angle plate.

Those with which or along which scribed lines are made

In this example a scribing block (surface gauge) is being used to transfer the dimension from the rule and to scribe a line parallel to the datum surface (and thus the datum edge of the component) at the required distance from the datum surface.

Where greater precision is required the vernier height gauge is used, as shown in Figure 5.7. Slip gauges and slip gauge accessories can also be used where ultimate precision is required, as shown in Figure 5.8.

Cylindrical components

Cylindrical components require special consideration. To prevent such components rolling about they are usually supported on V-blocks or on rollers, as shown in Figure 5.9. Such supports are usually sold in matched pairs for the support of long work parallel to the datum surface; they must be kept as matched pairs and not mixed with other, similar equipment. The plain straight edge is not

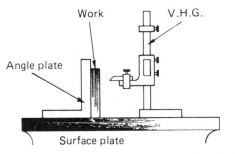

Figure 5.7 Marking out with the vernier height gauge

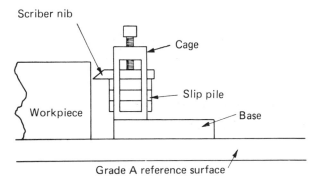

Figure 5.8 Marking out using slip gauges and accessories

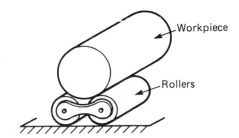

(a) USE OF ROLLERS TO SUPPORT A CYLINDRICAL COMPONENT

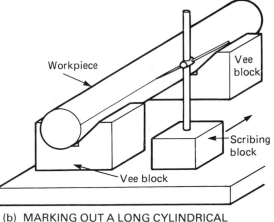

(b) MARKING OUT A LONG CYLINDRICAL COMPONENT

Figure 5.9 Marking out a long cylindrical component

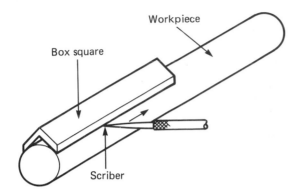

Figure 5.10 Use of the box square

suitable for marking lines parallel to the axis of cylindrical components, and a box square should be used as shown in Figure 5.10.

5.4 Materials for marking-out equipment

Cast iron   This is used for larger equipment such as surface tables, angle plates, parallel blocks, and large V-blocks. Its suitability was discussed in Section 4.11.

Granite   This is also used for large surface tables. Like cast iron, it is a dense, stable material. Compared with cast iron it has the advantage of not throwing up a burr if the working surface is scratched. However, it is softer and more susceptible to wear. It is not an anti-friction material, and marking-out tools will not slide so easily on its surface as they will on cast iron.

Hardened steel   Hardened and tempered high-carbon steel is used for all scribing tools which have to maintain a sharp point or edge. Parallel strips are also made from this material, as are rules, straight edges and try-squares. Steel will not slide smoothly over steel without a lubricant, but steel slides smoothly over cast iron. Medium-carbon steel and tool steel is used for centre punches and dot punches.

Carbide   Metal carbides are increasingly used for scribing points and edges. They are extremely hard-wearing materials, and tools maintain their cutting points or edges for long periods without resharpening. Carbides are brittle, and tools must

be used with great care to avoid chipping the points or edges.

5.5 Datum points, lines and surfaces

Throughout this chapter examples have been shown of some of the techniques for marking out straight lines, circles and arcs, lines parallel or perpendicular to a surface table, and lines along shafts. It is now necessary to consider the use of datum points, lines and surfaces.

Point datum   A single point from which a number of features are marked out: for example, concentric circles forming the inside and outside diameters of a flange ring, and the pitch circle around which a number of holes are stepped off (Figure 5.11a).

Line datum   A line from which or along which a number of features are marked out (Figure 5.11b).

Surface datum   Also known as an edge datum and a service edge. This is the most widely used datum for marking out. One edge or, more usually, two edges at right angles to each other (mutually perpendicular) are prepared on the component blank, and all dimensions for marking out are taken from these edges. If marking out takes place on a surface table, then these edges rest directly on the table or on suitable parallel strips to lift the blank to a convenient height. The positions of holes or other features are then located by means of polar or rectangular coordinates, as shown in Figure 5.11c.

5.6 Use of the coordinate table

Positioning holes by centre punching and drilling is not always sufficiently accurate, and it is sometimes necessary to use a coordinate drilling and boring machine such as a jig boring machine, a vertical milling machine, a computer numerically controlled (CNC) machining centre, or even a drilling machine fitted with a coordinate table, as shown in Figure 5.12. Longitudinal movement of the table is represented by the X axis and transverse movement by the Y axis, and the vertical movement of the spindle is represented by the Z

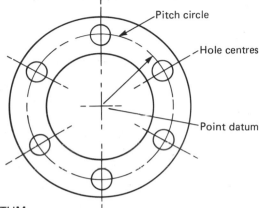

(a) SINGLE-POINT DATUM

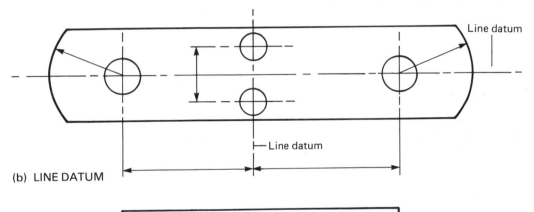

(b) LINE DATUM

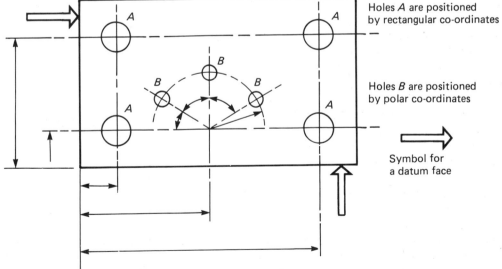

(c) SURFACE DATUM

Figure 5.11 Datums for marking out

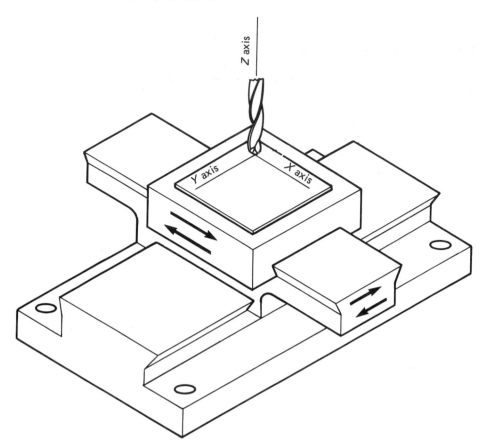

Figure 5.12 Coordinate drilling table: micrometer adjustment of X and Y slide omitted for clarity

axis. Figure 5.13 shows a typical component and the operation sheet for drilling the holes in blanks with previously machined edges. This avoids the need for marking out. Since all drilled holes tend to wander, the holes should preferably be finished by single-point boring.

5.7 Efficient marking out

Care of equipment

Like any other precision engineering equipment and tools, marking-out equipment must be carefully used and maintained if it is to work properly:

(a) The points of scribing instruments must be regularly dressed with an oil slip stone to keep them needle sharp. Letting them get so blunt that grinding has to be resorted to is bad practice, and a point produced on a grinding machine rarely has a satisfactory profile. Further, the heat generated by grinding is not conducted away quickly enough by the slender scriber, resulting in the point over-heating and softening as its temper is drawn (Figure 5.14a).

(b) Centre and dot punches should be kept sharp by grinding so that the grinding marks are parallel to the axis of the punch, as shown in Figure 5.14b.

(c) Vernier height gauge scribing blades should be ground as shown in Figure 5.14c, so that the datum edge is not destroyed.

(d) All items of precision equipment should be

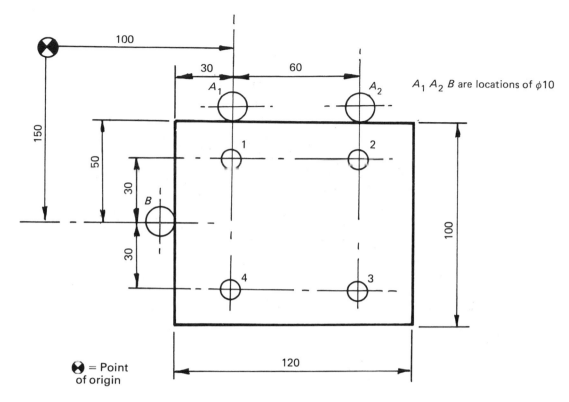

A_1 A_2 B are locations of $\phi 10$

⊗ = Point of origin

DIMENSIONS IN MILLIMETRES

Item	x	y	Sequence
⊗	100.00	150.00	
A_1	100.00	95.00	Setting
B	65.00	150.00	
	—	—	Load
Hole 1	100.00	120.00	Drill $\phi 5$
Hole 2	160.00	120.00	Drill $\phi 5$
Hole 3	160.00	180.00	Drill $\phi 5$
Hole 4	100.00	180.00	Drill $\phi 5$
	—	—	Remove

Figure 5.13 Operation sheet for coordinate drilling

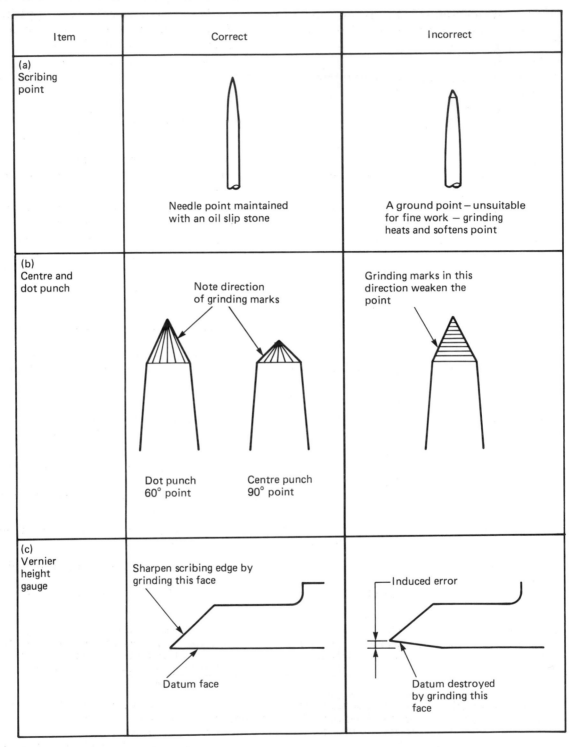

Item	Correct	Incorrect
(a) Scribing point	Needle point maintained with an oil slip stone	A ground point – unsuitable for fine work – grinding heats and softens point
(b) Centre and dot punch	Note direction of grinding marks Dot punch 60° point Centre punch 90° point	Grinding marks in this direction weaken the point
(c) Vernier height gauge	Sharpen scribing edge by grinding this face Datum face	Induced error Datum destroyed by grinding this face

Figure 5.14 Care of scribing points

cleaned and returned to their cases after use for safe storage and avoidance of corrosion.

Productivity

As in any other engineering operation, productivity when marking out can be increased by forethought and planning:

(a) Before starting, make sure that all the equipment required is readily to hand and that it is in good condition.

(b) Plan the sequence of marking-out operations so that the minimum number of moves have to be made. For example, all parallel lines on the same face should be made at the same setting of the workpiece. Scribe the minimum number of lines required.

(c) Make sure that the work is securely mounted so that it does not slip or deflect and that it is easily accessible from the working position.

(d) Ensure that the working position is adequately lighted without shadows and that the instruments can be clearly read.

Safety

As in any other engineering operation, safety precautions are important when marking out:

(a) Make sure heavy workpieces are securely mounted so that they cannot fall from the surface table, and that any mounting equipment (angle plates etc.) is capable of safely carrying the weight of the work.

(b) Heavy work should be lifted on to the surface table using a hoist so that the table is not damaged and the craftsperson is not strained by manual lifting.

Table 5.1 Faults and inaccuracies when marking out

Fault	Possible cause	Correction
Inaccurate measurement	1 Wrong instrument for tolerance required 2 Incorrect use of instrument	1 Check instrument is suitable for tolerance 2 Improve technique
Scribed lines out of position	1 Parallax (sighting) error 2 Rule not square with datum edge	1 Use scriber correctly 2 Use a datum block (abutment)
Lines not clear	1 Scribing point blunt 2 Work surface too hard 3 Scribing tool lacks rigidity	1 Sharpen point 2 Use a surface coating (spray-on lacquer) 3 Use only good quality tools in good condition
Corrosion along scribed lines	Protective coating (tin plate) cut by too sharp scribing point	Use a pencil when marking coated materials
Component tears or cracks along scribed line when bent	1 Scribed line and direction of bend parallel to grain of material 2 Scribed line cut too deeply	1 Bend at right angles to grain of material 2 Mark bend lines with a pencil
Circles and arcs irregular and not clear	1 Scribing points blunt 2 Instruments not rigid 3 Centre point slipping	1 Resharpen 2 Use only good quality dividers or trammels of correct size for job 3 Use dot punch to make centre location
Centre punch marks out of position	1 Incorrect use of punch 2 Scribed lines not sufficiently deep to provide point location	1 Position punch so that point is visible and then move upright when point is correctly positioned 2 Ensure point can click into junction of scribed lines

(c) Make sure that all loose equipment is positioned so that it cannot fall from the surface table on to the feet of the craftsperson. Tall and unstable equipment such as vernier height gauges should be laid on their sides when not being used so that they are not inadvertently knocked over and damaged.

(d) Sharp scribing points should be protected when not in use by means of a small cork, and at no time should unprotected sharp-pointed instruments be carried loose in an overall pocket.

Faults and inaccuracies

Some of the more common faults and inaccuracies which can occur during marking out are listed in Table 5.1, together with possible causes and corrections.

Exercises

For each exercise, select *one* of the four alternatives.

1 Scribed lines parallel to a datum edge are best made with
 (a) external calipers
 (b) hermaphrodite (odd-leg) calipers
 (c) scriber and try-square
 (d) dividers.
2 Scribed lines perpendicular to a datum edge are best made with
 (a) internal calipers
 (b) hermaphrodite (odd-leg) calipers
 (c) scriber and try-square
 (d) dividers.
3 A scribed line is best protected by
 (a) making a series of dot punch marks along it
 (b) spraying it with lacquer
 (c) coating it with copper sulphate solution
 (d) making a series of heavy centre punch marks along it.
4 Circles and arcs are usually marked out using
 (a) scriber and template
 (b) hermaphrodite (odd-leg) calipers
 (c) compasses
 (d) dividers.

5 A box square is used when
 (a) scribing lines parallel to the axis of a cylindrical component
 (b) marking out sheet metal for making bins and boxes
 (c) scribing lines around the circumference of a cylindrical component
 (d) folding the sides of sheet metal bins or boxes square with each other.
6 The holes shown in Figure 5.15 have been marked out using
 (a) polar coordinates
 (b) rectangular coordinates
 (c) circular coordinates
 (d) perpendicular coordinates.

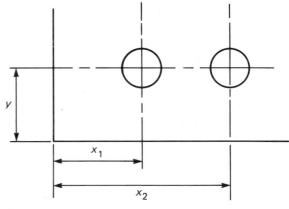

Figure 5.15

7 The holes shown in Figure 5.15 have been marked out from
 (a) centre line datums
 (b) edge (surface) datums
 (c) perpendicular datums
 (d) point datums.
8 The parallel lines being marked out with a vernier height gauge in Figure 5.16 are distance D apart. This is equal to height gauge settings of
 (a) $D_1 - D_2$
 (b) $D_1 + D_2$
 (c) $D_2 - D_1$
 (d) $D_2 + D_1$.
9 A coordinate table is normally used for positioning components for

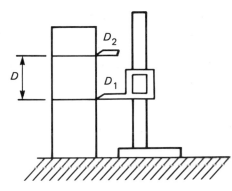

Figure 5.16

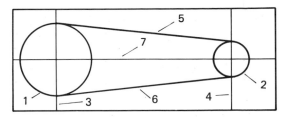

Figure 5.17

(a) drilling or boring holes without the need for marking out
(b) plotting polar coordinates
(c) marking out rectangular coordinates
(d) positioning components after the coordinates have been marked out.

10 A surface gauge (scribing block) is used mainly for
(a) scribing lines perpendicular to a datum surface
(b) scribing lines parallel to a datum surface
(c) scribing circles and arcs
(d) checking surface finish ready for marking out

11 One of the purposes of marking out is to provide
(a) guide lines to work to when only limited accuracy is required
(b) identification marks for use during assembly
(c) a means of controlling the size of a component where very high levels of accuracy are required
(d) a very accurate means of locating a drill point.

12 A suitable sequence for marking out the component shown in Figure 5.17 could be
(a) 1,2,3,4,5,6,7
(b) 7,3,4,1,2,5,6
(c) 7,3,4,5,6,1,2
(d) 1,2,5,6,7,3,4.

13 The fold lines on a tin plate blank should be marked out with a soft lead pencil
(a) because it shows up better than a scribed line
(b) because a drawn line is more permanent than a scribed line
(c) because it results in a more accurate component
(d) to avoid cutting through the tin plating so that corrosion occurs.

14 The device shown in Figure 5.18 is used when marking out
(a) mitred corners
(b) triangles
(c) holes from a datum edge
(d) the centre lines of circular blanks so as to determine the centre point.

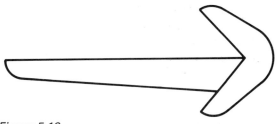

Figure 5.18

15 A major disadvantage of marking out with a sharp steel scriber is that
(a) the surface of the work is defaced
(b) the lines are easily removed
(c) the lines are thick and indistinct
(d) it will not mark steel workpieces.

6

Workholding and toolholding

6.1 The need for workholding: location and restraint

Figure 6.1 shows a metal block clamped to the table of a drilling machine ready for a hole to be drilled into it. In order for this operation to proceed correctly, a number of basic objectives have to be achieved:

(a) The block must be *located* under the drill so that the hole will be drilled in the correct place.
(b) The table of the drilling machine must *restrain* the block so that the feed force F_f of the drill does not push the component downwards. This is called *positive restraint*, since the workpiece is in contact with and restrained by a solid immovable object.
(c) The block has to be restrained so that it cannot be rotated by the drill. In this example clamps are used, and these apply *frictional restraint*. Although adequate for light machining operations, it is preferable to use some form of positive restraint to resist the main cutting forces F_c.

Thus workholding consists of a combination of positional location together with positive and/or frictional restraint to maintain the workpiece in its correct position and prevent it being displaced by the cutting forces and its own weight. The principles of location and restraint will now be considered in greater detail.

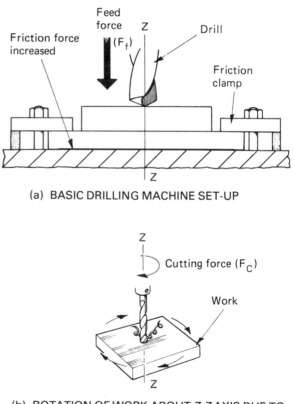

(a) BASIC DRILLING MACHINE SET-UP

(b) ROTATION OF WORK ABOUT Z-Z AXIS DUE TO DRILLING TORQUE

Figure 6.1 Need for workholding

6.2 Basic concepts of workholding

Figure 6.2 shows a body in space, free to move in any direction. It can be seen that the body is free to move along any of the three axes or rotate about those axes, and so has six degrees of freedom. In order to work on that body it must be *located* in a given position by *restraining* its freedom of movement.

Figure 6.3 shows the principle of location. The base or first plane surface contains three points of restraint for the component along the *Z-Z* axis, and about the axes *X-X* and *Y-Y*. The second plane contains two points of restraint which restrain the component along *Y-Y* and about *Z-Z*. Finally, the third plane provides one point of restraint, the component along *X-X*. Since this arrangement incorporates an effective clamping system, the six degrees of freedom are fully eliminated.

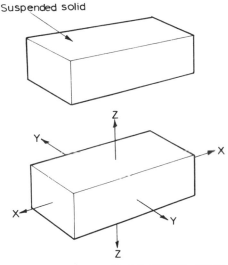

Suspended solid

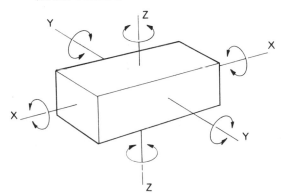

THREE FREEDOMS OF TRANSLATION

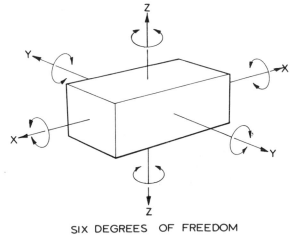

THREE FREEDOMS OF ROTATION

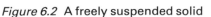

SIX DEGREES OF FREEDOM

Figure 6.2 A freely suspended solid

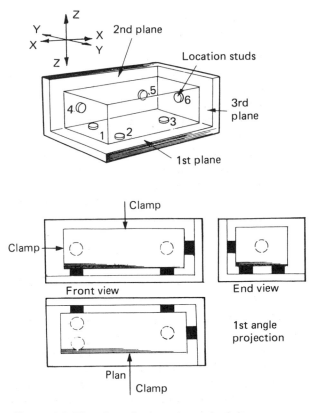

Figure 6.3 The six-point location principle

When one is designing a workholding device, the number of restraints should be kept to the minimum necessary, and solid rather than frictional restraints should be used to resist the major cutting forces. Positional location should always be achieved by the use of positive restraints.

6.3 Precautions when workholding

Distortion of the workpiece can occur if it is not properly supported to resist the forces applied by the clamps and the cutting tool, as shown in Figure 6.4.

Damage to the workpiece can also occur if it is not correctly supported or if the type of clamp selected is incorrect. Figure 6.5a shows how thin hollow castings can be broken by incorrect clamping. Again, the type of clamp selected depends upon the type of surface it is to act upon. For example, Figure 6.5b shows fibre shoes for preventing the serrations on the jaws of a fitter's bench vice from marking machined and ground surfaces. Similarly, previously machined surfaces must be protected as shown in Figure 6.5c.

Support of the workpiece as closely as possible to the point of cutting is essential. This can be a concept of design as shown in Figure 6.6a, or of setting as shown in Figures 6.6b and 6.6c.

Cylindrical workpieces present special problems. Figure 6.7 shows how a V-block is used to locate cylindrical components in the horizontal and the vertical plane. In addition, round bars and components can be held in various types of chuck as used on lathes, grinding machines and dividing heads.

6.4 Workholding applications

Vices

These are quick and convenient devices for holding work of a variety of shapes and sizes.

Parallel-jaw bench vice (Figure 6.8a) The component being held should be positioned so that, wherever possible, the main cutting or bending force being applied to the workpiece acts towards the fixed jaw.

Pipe vice (Figure 6.8b) This is used for holding cylindrical bars or tubes, mainly for threading operations. The jaws, whose V formation makes them self-centring, are heavily serrated to obtain a good grip on the work.

Parallel-jaw machine vice (Figure 6.8c) This is bolted to the machine worktable and holds the work rigidly against the cutting action of the machine. Unlike the bench vice, the jaws of the machine vice are ground smooth so that they do not mark the work.

Clamps

These are used for holding work directly on to the machine table. They are used for large work and where heavy cuts are to be taken. Figure 6.9a

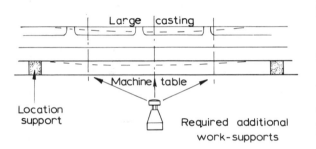

(a) HEAVY COMPONENT DEFLECTING ON ITS TWO WIDELY SPACED LOCATIONS

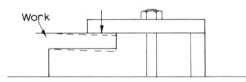

(i) Incorrect clamping, work distorts

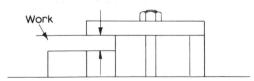

(ii) Correct clamping, work supported

(b) THE CLAMPING OF THIN-SECTION WORK

Figure 6.4 Precautions when workholding

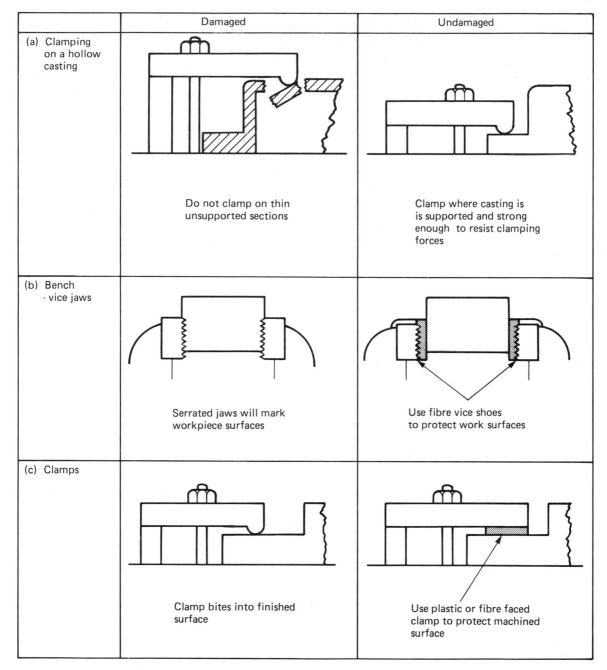

	Damaged	Undamaged
(a) Clamping on a hollow casting	Do not clamp on thin unsupported sections	Clamp where casting is is supported and strong enough to resist clamping forces
(b) Bench - vice jaws	Serrated jaws will mark workpiece surfaces	Use fibre vice shoes to protect work surfaces
(c) Clamps	Clamp bites into finished surface	Use plastic or fibre faced clamp to protect machined surface

Figure 6.5 Avoiding damage when clamping

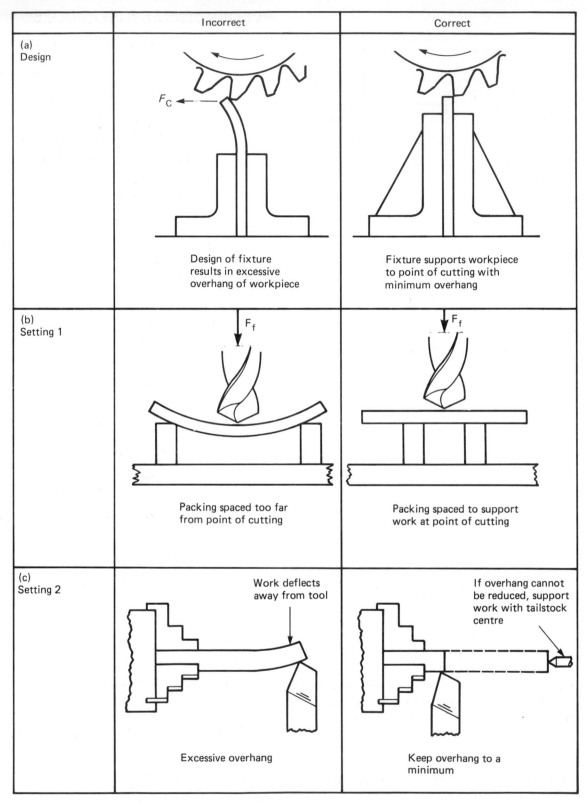

	Incorrect	Correct
(a) Design	Design of fixture results in excessive overhang of workpiece	Fixture supports workpiece to point of cutting with minimum overhang
(b) Setting 1	Packing spaced too far from point of cutting	Packing spaced to support work at point of cutting
(c) Setting 2	Excessive overhang	Keep overhang to a minimum

Figure 6.6 Supporting the workpiece

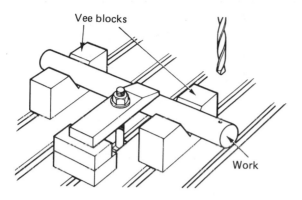

(a) HORIZONTAL

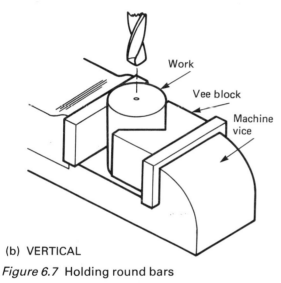

(b) VERTICAL

Figure 6.7 Holding round bars

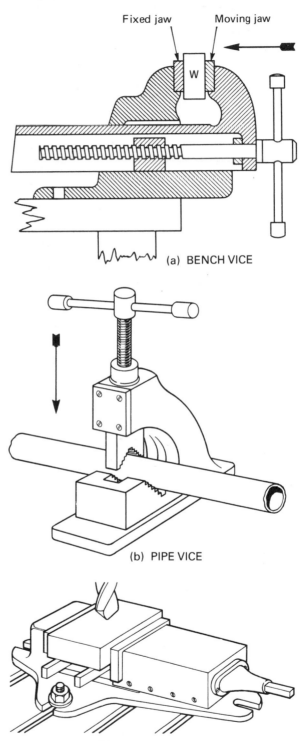

(a) BENCH VICE

(b) PIPE VICE

(c) MACHINE VICE

Figure 6.8 Vices

shows how clamps are used to prevent movement of the work during machining. The height of the packing should be the same as the height of the work, with the bolt nearer the work than the packing. The clamps force the work down against the machine table, and the friction between the work and the machine table provides the restraint against the cutting force. Figures 6.9b and 6.9c show some different types of clamps and the principles of clamping.

If the cutting force is large, then a positive restraint in the form of a fixed stop is required, as shown in Figure 6.9d. Fixed stops should be positioned so that the main cutting force is towards

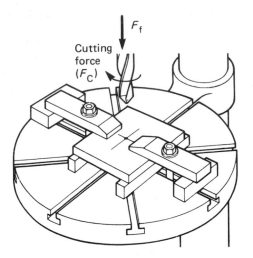

(a) WORK SUPPORTED ON PARALLELS AND
CLAMPED TO TABLE

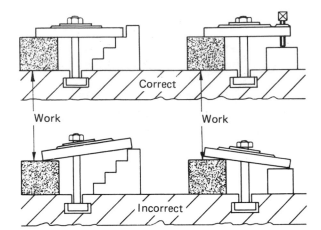

(c) CLAMPING PRINCIPLES

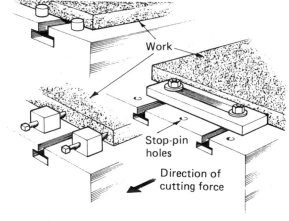

(b) THREE USEFUL CLAMPS

(d) POSITIVE END STOPS

Figure 6.9 Use of clamps

them. The clamping of cylindrical work was shown in Figure 6.7a.

Angle plate

In all the previous examples the hole to be drilled was *perpendicular* to the datum surface. Figure 6.10 shows the use of an angle plate to support the work when the hole to be drilled is *parallel* to the datum surface (base of the workpiece). A further example is shown in Section 6.5 for the face-plate.

6.5 Lathe workholding

Machine axes

When the six degrees of freedom of a body in space were introduced, the Z axis was placed in the vertical plane. It has been continued to be shown in this position in the subsequent examples, which have been based on the use of the drilling machine. This is correct because the appropriate British Standard states that the main spindle axis of any

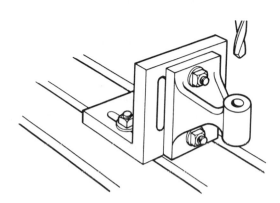

Figure 6.10 Use of angle plate

drilled. It is then removed and the lathe is changed over for holding between centres. Grease should be placed in the centre holes if dead (stationary) centres are used in order to reduce friction and damage to the centres. The tailstock centre should be eased from time to time as the work warms up and expands.

Face plate

Figure 6.12c shows a component being held on a face plate so that the hole can be bored perpendicular to the datum surface which is in contact with the face plate. Note that the face plate has to be balanced to ensure smooth running when the work and any additional workholding device such as an angle plate is offset from the axis. Care must be taken when using the face plate that the clamps will hold the work securely so that it will not spin out of the lathe, and also that the clamps themselves will not foul any part of the lathe or its guards.

6.6 Toolholding applications

The principles of restraints and locations previously described for holding the workpiece apply equally to the holding of cutting tools.

Drills

Figure 6.13a shows a taper-shank drill mounted in a drilling machine spindle. The Morse taper system is universally employed for this purpose, as it is a self-locking taper and will locate and restrain the drill. That is, it will locate the drill so that it is in axial alignment with the spindle and will run true, and also it will hold the drill in position. The taper drives the drill and also prevents it from dropping out of the spindle. The tang on the end of the drill shank is only for extraction purposes, as shown in Figure 6.13b; it does *not* drive the drill. To ensure true running the taper shank of the drill and the bore of the machine spindle must be kept clean and free from burrs. Adaptor sleeves are used where the shank taper is smaller than the spindle taper.

The drill chuck is used for holding plain-shank drills. As the outer sleeve is rotated by the chuck key, the jaws are driven down against a taper in the chuck body and close concentrically upon the drill

machine is called the Z axis, no matter whether this is the axis about which the work rotates or the axis about which the tool rotates. Thus in a drilling machine the drill spindle axis is vertical, so the Z axis is also vertical. However, in the lathe the spindle axis and therefore the work axis are horizontal, so the Z axis for a lathe is also horizontal. The X and Y axes retain their relationship with the Z axis, as shown in Figure 6.11. The X axis is always horizontal and parallel to the workholding surface. The Y axis is always perpendicular to both the Z and the X axes.

Self-centring chuck

Figure 6.12a shows a three-jaw self-centring chuck. The jaws are set 120° apart and move in and out simultaneously (at the same time) when the key is rotated. This key *must* be removed before the machine is set in motion, or a serious accident will occur. The movement of the jaws ensures that hexagonal or cylindrical work will always be held with its axis concentric with the axis of the chuck: hence the name 'self-centring chuck'. The restraints and locations acting upon the workpiece when it is held in such a chuck are also shown.

Between centres

Figure 6.12b shows how work to be turned in the lathe is held between centres and driven by a carrier and catchplate. The restraints and locations for work held between centres are also shown. The work is first held in a self-centring chuck so that it can be faced at each end and the centre holes

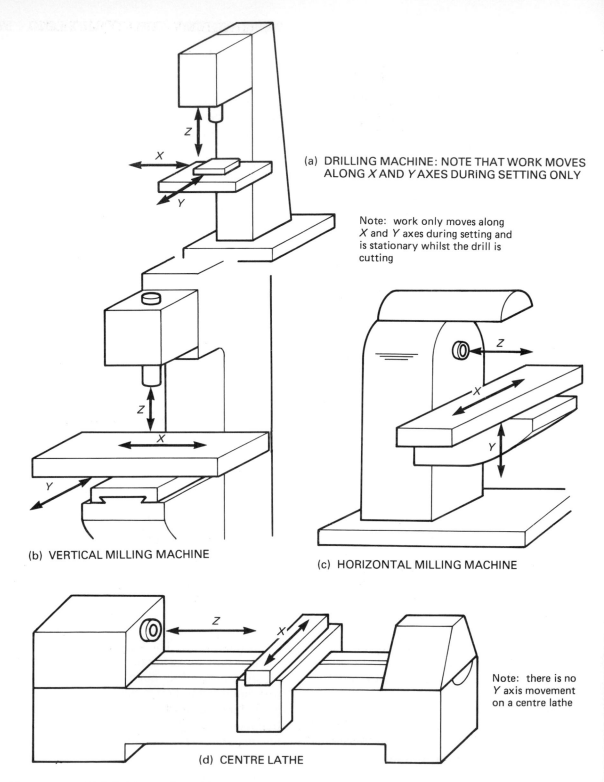

(a) DRILLING MACHINE: NOTE THAT WORK MOVES ALONG *X* AND *Y* AXES DURING SETTING ONLY

Note: work only moves along *X* and *Y* axes during setting and is stationary whilst the drill is cutting

(b) VERTICAL MILLING MACHINE

(c) HORIZONTAL MILLING MACHINE

Note: there is no *Y* axis movement on a centre lathe

(d) CENTRE LATHE

Figure 6.11 BSI definitions of machine axes

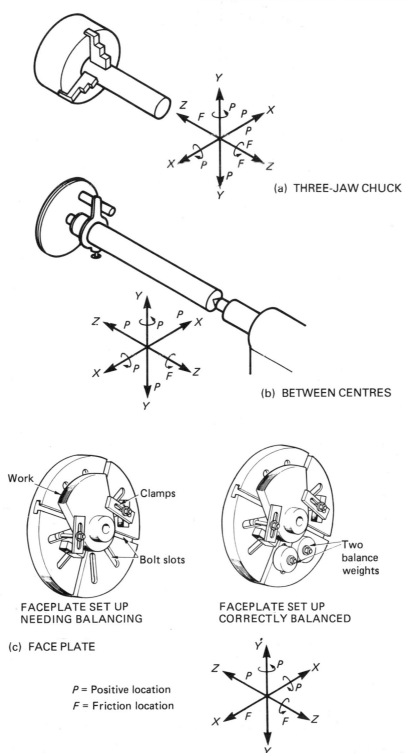

(a) THREE-JAW CHUCK

(b) BETWEEN CENTRES

Work

Clamps

Bolt slots

Two balance weights

FACEPLATE SET UP
NEEDING BALANCING

FACEPLATE SET UP
CORRECTLY BALANCED

(c) FACE PLATE

P = Positive location
F = Friction location

Figure 6.12 Workholding on the lathe

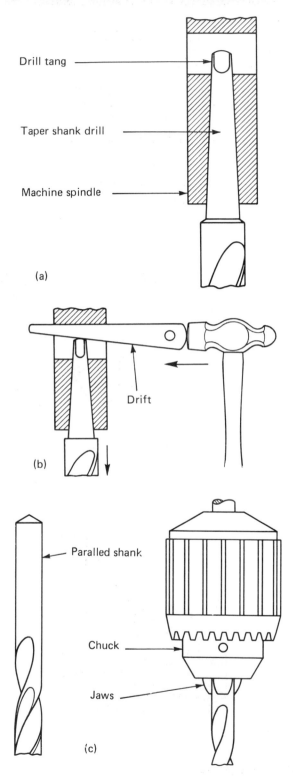

shank. The chuck key *must* be removed before starting the machine, or a serious accident will occur. Drill chucks are generally used for the smaller sizes of drill. An example of a drill chuck is shown in Figure 6.13c.

Milling cutters

These are mounted on arbors or are held in collet chucks. Figure 6.14a and b shows a horizontal milling machine arbor mounted in the spindle nose of a milling machine and carrying a typical milling cutter. Unlike the drilling machine, the spindle nose of the milling machine has a quick-release taper (MMT series). This taper merely locates the arbor which is restrained by a drawbolt passing through the machine spindle. The arbor is driven by dogs on the spindle nose. The milling cutter has a parallel bore and is a close slide fit on the arbor. Any play at this point would result in the cutter running off-centre. The milling cutter is positioned on the arbor by the use of spacing collars and is driven by a key as shown.

Stub arbors of various types are also used, and these may carry face milling cutters as shown in Figure 6.14c.

End mills and slot drills are held in collet chucks as shown in Figure 6.14d. The collet is closed on the shank of the cutter by a system of concentric tapers to ensure axial alignment and true running.

Single-point tools

These have a rectangular shank and are clamped into the toolpost. Figure 6.15a shows a four-way turret-type toolpost. The tools have to be set to centre height by the use of packing strips and adjusted for position by the operator. However, once this has been done they can be quickly swung into position as required, and this saves time on batch production. Figure 6.15b shows a single-point tool set in the toolpost of a shaping machine.

Figure 6.13 (a) Holding taper-shank drill (b) extracting taper-shank drill (c) holding parallel-shank drill

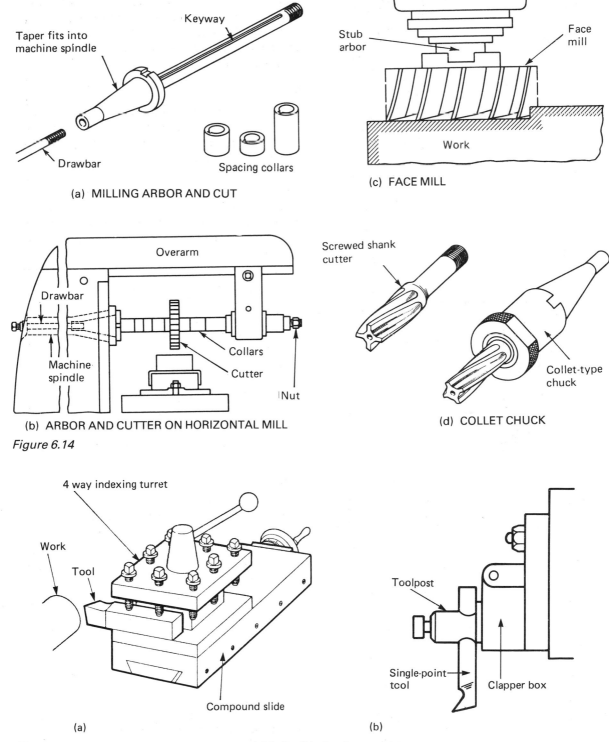

(a) MILLING ARBOR AND CUT

(c) FACE MILL

(b) ARBOR AND CUTTER ON HORIZONTAL MILL

(d) COLLET CHUCK

Figure 6.14

(a)

(b)

Figure 6.15 Holding single-point tools on (a) lathe (b) shaping machine

Exercises

For each exercise, select *one* of the four alternatives.

1 The main cutting force acting on the workpiece should
   (a) be resisted by a frictional restraint
   (b) be resisted by a positive restraint
   (c) not be resisted to avoid deforming the workpiece
   (d) not be resisted to reduce the load on the cutting tool.

2 In which one of (a), (b), (c) or (d) in Figure 6.16 are the six degrees of freedom fully restrained?

3 When work is held in a vice, the main cutting force should be directed
   (a) towards the fixed jaw
   (b) towards the moving jaw
   (c) parallel to the fixed jaw
   (d) always vertically downwards.

4 Fibre shoes are sometimes fitted over the jaws of a fitter's bench vice to
   (a) increase the friction between the jaws and the work
   (b) reduce the friction between the jaws and the work
   (c) avoid marking the job
   (d) avoid damaging the jaws.

5 Which one of (a), (b), (c) or (d) in Figure 6.17 shows the clamp correctly fitted?

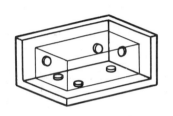
(a)

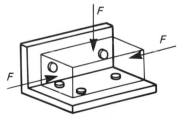
(b)

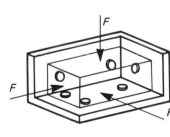
(c)

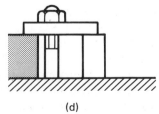
(d)

Figure 6.16

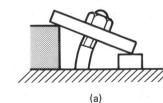
(a)

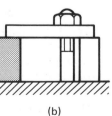
(b)

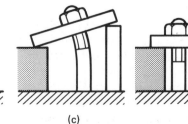
(c)

(d)

Figure 6.17

6 To drill the hole through the component shown in Figure 6.18, perpendicular to surface A and parallel to surface B, the component should be held
   (a) in a machine vice with surface A against the fixed jaw
   (b) by clamping it to the drilling machine table
   (c) by hand because it is an awkward shape
   (d) by clamping it to an angle plate

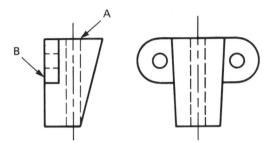

Figure 6.18

7 In a three-jaw self-centring chuck as used on a centre lathe, small-diameter work is driven against
   (a) both the main cutting force and the feed force by frictional restraints only
   (b) the main cutting force by positive restraint and the feed force by frictional restraint
   (c) the main cutting force by frictional restraint and the feed force by positive restraint
   (d) both the main cutting force and the feed force by positive restraint.

8 Work held between centres should be measured at both ends after the first cut to ensure
   (a) it is the correct size
   (b) it is parallel
   (c) it is not oval
   (d) it is the correct length

9 When roughing out long work supported at one end on a revolving centre fitted in the tailstock, the centre should be eased from time to time to
   (a) allow for expansion as the work heats up
   (b) inspect the condition of the centres
   (c) inspect the condition of the centre hole
   (d) lubricate the centre.

10 To ensure concentricity between the various diameters of a stepped component,
   (a) it should be reset after each diameter is turned
   (b) it should be trued up with a DTI when it is first put in the chuck
   (c) all the diameters should be turned at the same setting
   (d) it should only be turned between centres.

11 To rebore the diameter D of the bearing housing shown in Figure 6.19 so that the axis of the bore is truly perpendicular to the surface A, the housing is best held in a lathe
   (a) on a face plate
   (b) between centres
   (c) in a three-jaw self-centring chuck
   (d) on an angle plate.

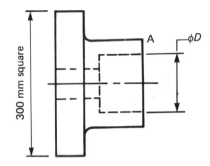

Figure 6.19

12 Since the taper bore of a drilling machine spindle needs to be self-locking, it is usually
   (a) a Morse taper
   (b) an MMT taper
   (c) a Brown and Sharpe taper
   (d) specially designed by the machine maker.

13 Large taper-shank drills are driven
   (a) positively by the tang on the end of the taper
   (b) frictionally by the taper shank
   (c) by a combination of both (a) and (b)
   (d) positively by a drift inserted through the slot in the spindle.

14 Milling cutter arbors are usually
   (a) restrained by a drawbar and located by a Morse taper
   (b) driven by dogs on the spindle nose and located by a Morse taper

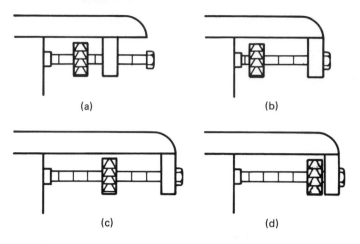

Figure 6.20

(c) driven by friction if an MMT series taper is used
(d) driven by dogs on the spindle nose and located by an MMT series taper.

15 Which one of the milling cutter settings (a), (b), (c) or (d) in figure 6.20 is the *least* desirable?

16 End mills and slot drills are usually held
(a) on a long arbor
(b) on a stub arbor
(c) directly on the spindle nose
(d) in a collet chuck.

17 Single-point lathe tools are usually held
(a) on an arbor
(b) in a chuck
(c) in a toolpost
(d) by clamping them directly to the cross-slide

18 Taper-shank drills should be removed from the drilling machine spindle by means of
(a) a taper drift acting on the tang of the drill
(b) an extractor acting on the spindle nose
(c) a rod passing through the spindle
(d) a soft-faced mallet used directly on the drill.

19 When drilling, the work should be securely clamped
(a) to prevent distortion of the workpiece
(b) to avoid injury to the operator
(c) to avoid damage to the machine and drill should the drill 'grab'
(d) for both the reasons given in (b) and (c)

20 Which one of the work settings (a), (b), (c) or (d) in Figure 6.21 is the most desirable? Note: F indicates the position of the clamping force.

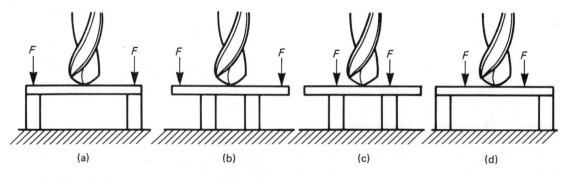

F = Clamping force

Figure 6.21

7

Material removal

7.1 The cutting edge

Perhaps the first controlled cutting action any of us made was the sharpening of a pencil with a penknife. It is highly unlikely that we received any formal instructions before our first atttempt but, by trial and error, we found that the blade of the knife had to be inclined to the wood at a definite angle before cutting took place. This established the need for a *clearance angle*. It was also found that the knife cut more easily (cutting force was less) if the cutting edge was sharp.

If the wooden pencil is replaced by a piece of soft metal rod such as brass or aluminium it will be found that the blade quickly becomes blunt and, if it is examined with the aid of a magnifying glass, it will be seen that the edge has crumbled away. For the blade to cut metal successfully, the cutting edge has to be sharpened to a less acute angle to give it greater strength. The angle to which the blade is ground is called the *wedge angle*.

Figure 7.1 shows the angles introduced above applied to a single-point tool such as a shaping machine tool. It can be seen that there is one more angle to be considered – the *rake angle*. This is a very important angle as it controls the cutting action of the tool. For most ductile materials, the larger the rake angle can be made, the more easily the tool will cut. However it can be seen that if the rake angle is made larger, the wedge angle becomes smaller and the tool becomes weaker. Therefore the geometry of the tool is a compromise between cutting efficiency and tool

Material	Rake angle
Aluminium alloy	30°
Brass (ductile)	14°
Brass (free-cutting)	0°
Cast iron	0°
Copper	20°
Phosphor bronze	8°
Mild steel	25°
Medium carbon steel	15°

For high speed steel tools under normal workshop conditions

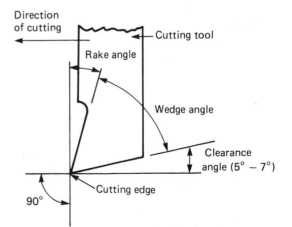

Direction of cutting

Cutting tool

Rake angle

Wedge angle

Clearance angle (5° – 7°)

Cutting edge

90°

Figure 7.1 Definitions of cutting angles
Rake angles for high speed steel tools under normal workshop conditions.

strength. Figure 7.1 also lists some typical values of clearance and rake angles for typical operations and materials when using high-speed steel tools. If the clearance angle is made too small the tool will rub, become hot and soften; it will then quickly become blunt, and may even be destroyed. If the clearance angle is made too large, not only is the wedge angle reduced and the tool weakened but also the tool will tend to dig in and chatter, leaving a poor finish.

When cutting high-strength alloys with very hard but brittle cutting tool materials, such as carbides or ceramics, it is necessary to give the cutting edge *negative rake*. This not only increases the strength of the tool tip, but raises the temperature of the workpiece at the point of cutting to such an extent that the workpiece material becomes more ductile and its strength is reduced at this point, making it easier to cut. Very powerful and rigid machines are required if these conditions are to be achieved. Figure 7.2 compares positive, neutral and negative rake as applied to a single-point tool.

The metal removed by the tool is called the *chip*. This varies from granules in the case of cast iron and free-cutting brass, which are relatively brittle and have a low ductility, to almost continuous ribbons of metal for highly ductile materials. This accounts for zero rake angles being listed in Figure 7.1 for cast iron and free-cutting brass. Not only does this strengthen the tool, but also the granular chips do not need to flow over the rake face of the

tool like the continuous chip of a ductile material. The chips removed by the cutting tool are also referred to as *swarf*.

Thus the factors which affect the penetration of the workpiece by the cutting edge may be summarized as:

Hardness of the cutting tool material. This must be harder than the workpiece so that it does not become blunt, but not so hard that the cutting edge is brittle, chips easily and loses its shape.

Sharpness of the cutting edge. A sharp cutting edge is required if a good surface finish is to be left on the workpiece and if the tool is to cut efficiently.

Wedge angle of the cutting tool. This must be adequate to give the required strength and tool life, but not so large as to reduce the rake angle and therefore the cutting efficiency of the tool.

7.2 The application of cutting angles

Consideration of the following cutting tools shows that they all have the basic rake, wedge and clearance angles described in Section 7.1. Practical considerations in the manufacture of the tool may modify the appearance of these angles and, in some circumstances, it is necessary to introduce additional angles.

Cold chisels

The application of cutting tool angles to this tool is shown in Figure 7.3. For cutting soft materials of low strength the wedge angle can be made more acute so that the chisel cuts more easily. For cutting harder materials and materials of greater strength the wedge angle is increased to give the chisel greater strength. Some typical examples of the wedge angle for various materials are listed in Figure 7.3, together with corresponding angles of inclination. The difference between half the wedge angle and the angle of inclination gives a constant clearance angle of 7°.

In practice, the fitter does not work out the angle of inclination or the rake and clearance angle, but uses experience and the feel of the chisel as it cuts through the metal to present it to the work at the correct angle. If the angle of inclination is

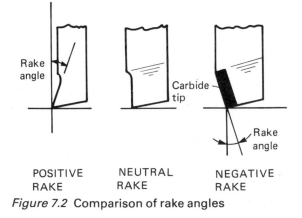

Figure 7.2 Comparison of rake angles

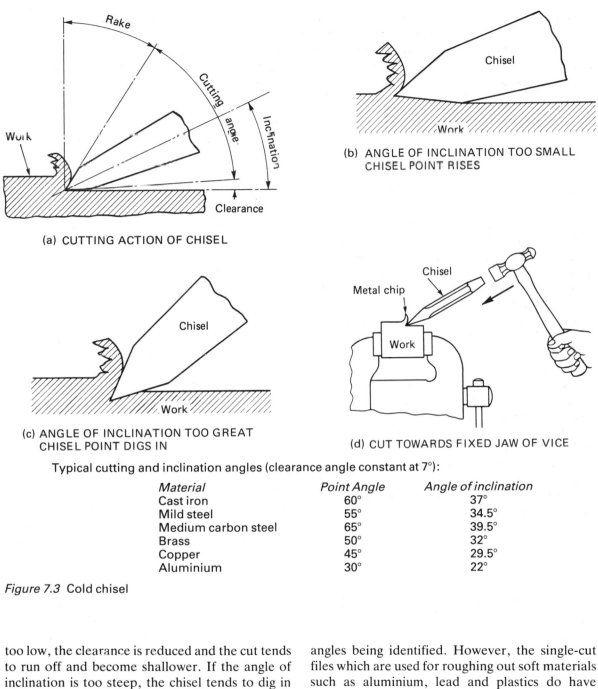

(a) CUTTING ACTION OF CHISEL

(b) ANGLE OF INCLINATION TOO SMALL
CHISEL POINT RISES

(c) ANGLE OF INCLINATION TOO GREAT
CHISEL POINT DIGS IN

(d) CUT TOWARDS FIXED JAW OF VICE

Typical cutting and inclination angles (clearance angle constant at 7°):

Material	Point Angle	Angle of inclination
Cast iron	60°	37°
Mild steel	55°	34.5°
Medium carbon steel	65°	39.5°
Brass	50°	32°
Copper	45°	29.5°
Aluminium	30°	22°

Figure 7.3 Cold chisel

too low, the clearance is reduced and the cut tends to run off and become shallower. If the angle of inclination is too steep, the chisel tends to dig in and the cut becomes progressively deeper.

Files

Most engineer's files are cross-cut or double-cut, but this type of cut prevents any definite cutting angles being identified. However, the single-cut files which are used for roughing out soft materials such as aluminium, lead and plastics do have clearly defined cutting angles, as shown in figure 7.4. The teeth may have positive or negative rake depending upon the process by which they were formed and the application for which the file is intended.

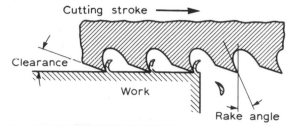

(a) CUTTING ACTION OF THE FILE

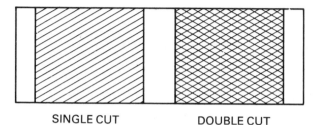

SINGLE CUT   DOUBLE CUT

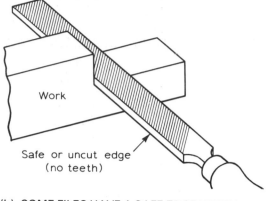

(b) SOME FILES HAVE A SAFE EDGE WHICH
    ALLOWS FILING TO TAKE PLACE WITHOUT
    DAMAGING ANOTHER FINISHED FACE

Figure 7.4 Engineer's file cutting action. File grades:

Cut	Use
Rough	Rapid removal of large amounts of material
Bastard	Rapid removal of smaller amounts of material
Second cut	Removing smaller amounts of material before finishing
Smooth	Smooth surface finish
Dead smooth	Special work with very smooth accurate finish

To preserve the cutting edges of the teeth the following precautions should be taken:

(a) New files should first be used on soft metals such as aluminium or copper.
(b) Never use a file on a metal of unknown hardness.
(c) Select the correct file for the job.
(d) Make sure the work is securely held with the minimum overhang so that it will not vibrate and chatter against the teeth.
(e) Never throw files into a drawer on top of each other. They should be racked up.
(f) Keep files clean and free from filings by regular brushing with a file card.

Hacksaw blades

Figure 7.5 shows the teeth of a hacksaw blade for use by hand. In order to provide adequate clearance, so that the chips may be carried out of the cut without clogging the saw, the clearance angle is exaggerated. The teeth of a hacksaw blade for use in a power hacksaw require greater strength and are given a normal clearance angle so as to provide a robust metal cutting wedge. This is called the primary clearance angle. In addition, a secondary clearance angle is provided to clear the chips from the cut.

Saw teeth are also given a *set* so that the slot cut is slightly wider than the thickness of the blade to prevent it jamming and breaking. In fine-tooth blades the whole of the blade is wave set, as the teeth are too fine to set individually. In coarser-tooth blades the teeth themselves are bent alternately to the left and to the right, with every fifth tooth left straight to clear the slot being cut.

For efficient cutting the following factors should be noted.

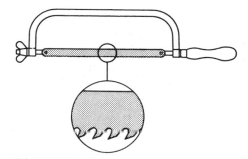

(a) HACKSAW

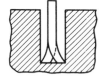

(b) THIN SECTIONS MORE TEETH

(c) SET OF TEETH

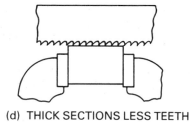

(d) THICK SECTIONS LESS TEETH

Figure 7.5 Hacksaw. Ensure that a minimum of three teeth are in contact with the work

Teeth per inch	Material to be cut	Blade application
32	up to 1/8 in	Thin sheets and tubes Hard and soft materials (thin sections)
24	1/8 to 1/4 in	Thicker sheets and tubes Hard and soft materials (thicker sections)
18	1/4 to 1/2 in	Heavier sections such as mild steel, cast iron, aluminium, brass, copper, bronze
14	1/2 in plus	Soft materials of heavy section such as aluminium, brass, copper, bronze

(a) High-speed steel blades have a longer life and should be used when cutting tough materials.
(b) Cheaper, low-speed tungsten blades are only suitable for soft materials such as brass, copper, aluminium and plastics.
(c) Do not overtighten the blade. Three turns of the wing nut should be sufficient once the slack has been taken up.
(d) Use long steady strokes, not short fast strokes.
(e) Do not start cutting on a corner.
(f) Ensure the work is securely clamped with the minimum overhang.

Shaping machine tools

The application of the basic cutting angles to a shaping machine tool has already been considered in Figure 7.1. Some shaping machine tool profiles are shown in Figure 7.6.

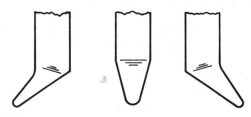

LEFT-HAND SHAPING TOOL   ROUND NOSE ROUGHING TOOL   RIGHT-HAND SHAPING TOOL

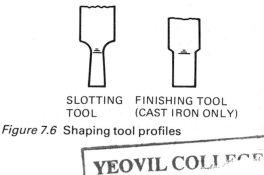

SLOTTING TOOL   FINISHING TOOL (CAST IRON ONLY)

Figure 7.6 Shaping tool profiles

YEOVIL COLLEGE

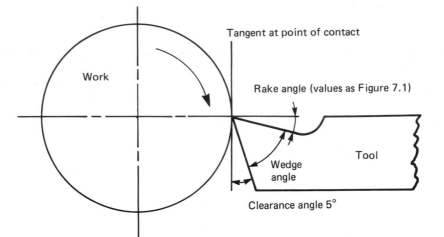

(a) TURNING TOOL SET CORRECTLY ON CENTRE (RAKE ANGLES AS FIGURE 7.1)

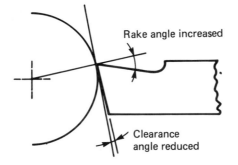

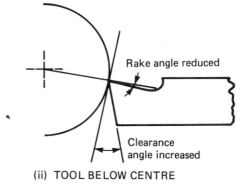

(i) TOOL ABOVE CENTRE

(ii) TOOL BELOW CENTRE

(b) TURNING TOOL SET INCORRECTLY

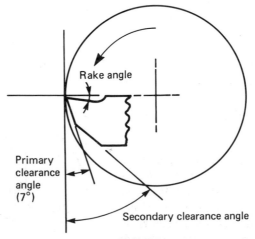

(c) BORING TOOL ANGLES

Figure 7.7 Effective tool height

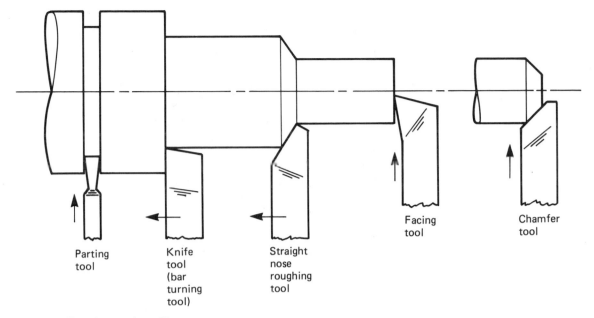

Figure 7.8 Turning tool profiles

Parting tool

Knife tool (bar turning tool)

Straight nose roughing tool

Facing tool

Chamfer tool

Turning tools

The application of the basic cutting angles to a turning tool is shown in figure 7.7a. Basically they are the same as for the shaping machine tool rotated through 90°. However, because the cutting edge is in contact with a curved surface the cutting angles are taken from imaginary planes at the point of contact as shown. Thus the cutting angles vary if the tool is mounted above or below the centre height of the workpiece, as shown in Figure 7.7b.

A boring tool has to have a secondary clearance angle to prevent the heel of the tool fouling the workpiece. This is shown in Figure 7.7c.

Figure 7.8 shows some turning tool profiles.

Twist drills

The application of the basic cutting angles to a twist drill is shown in Figure 7.9, where the drill is compared with a single-point cutting tool. It can be seen that the helix angle represents the rake angle at the outer edge of the lip of the drill. This angle is not constant and becomes less towards the centre of the drill, with a corresponding reduction in cutting efficiency.

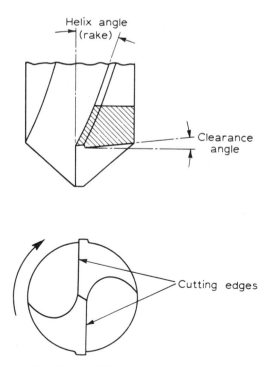

Figure 7.9 Twist drill cutting angles

Material to be Drilled	Helix angle HELIX ANGLE (RAKE)		POINT ANGLE
Aluminium	40°	Quick spiral	90°
Brass	0–10°	Slow spiral	118°
Cast iron	0–10°	Slow spiral	118°
Copper	30°	Quick spiral	90°
Hard steel	5–10°	Slow spiral	130°
Mild steel	25°	Standard spiral	118°
Plastics	25°	Standard spiral	90°

Figure 7.10 Twist drill point and helix angles

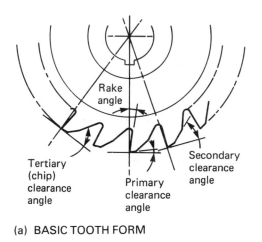

(a) BASIC TOOTH FORM

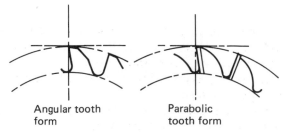

Angular tooth form   Parabolic tooth form

(b) ALTERNATIVE TOOTH FORMS

Figure 7.11 Milling cutter

It is not possible to vary the rake angle of a drill to any great extent during regrinding since the helix angle of the flutes is fixed at the time of manufacture. However, drills can be bought with various helix and point angles for drilling different materials, as shown in figure 7.10.

Milling cutters

Like the saw blade these are multitooth cutters, and chip clearance has to be provided to prevent clogging and breakage of the teeth. Figure 7.11a shows the cutting angles. Primary clearance is provided to give the essential metal cutting wedge. Secondary clearance is also provided to prevent the heel of the tooth from fouling the workpiece. A gullet (tertiary clearance) is then provided for chip clearance. Alternatively the tooth form may be as shown in Figure 7.11b. Only primary clearance is provided and the parabolic form of the tooth replaces the other angles.

7.3 Depth of cut and feed

Parallel turning

Figure 7.12a shows a parallel turning operation using an *orthogonal* turning tool: that is, a turning tool in which the cutting edge is at right angles to

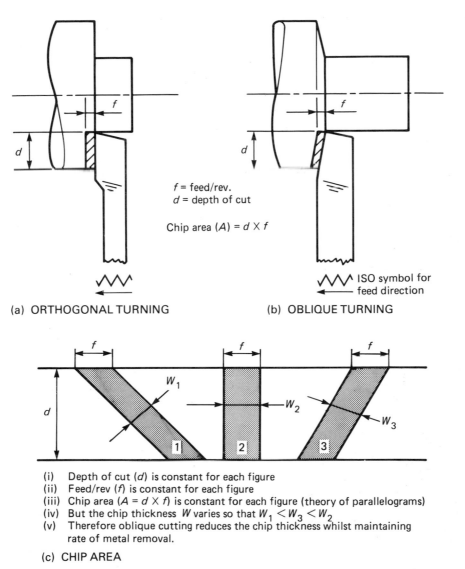

f = feed/rev.
d = depth of cut

Chip area (A) = d × f

ISO symbol for feed direction

(a) ORTHOGONAL TURNING   (b) OBLIQUE TURNING

(i)   Depth of cut (d) is constant for each figure
(ii) Feed/rev (f) is constant for each figure
(iii) Chip area (A = d × f) is constant for each figure (theory of parallelograms)
(iv) But the chip thickness W varies so that $W_1 < W_3 < W_2$
(v)   Therefore oblique cutting reduces the chip thickness whilst maintaining rate of metal removal.

(c) CHIP AREA

Figure 7.12 Feed and depth of cut for parallel turning

the direction of feed. The hatched rectangle represents the cross-sectional area df of the chip, where d is the depth of cut and f is the feed per revolution of the workpiece. The ratio of the depth of cut to feed is known as the $d{:}f$ ratio. Since the tool is cutting orthogonally and the cross-section of the chip is a rectangle, the chip thickness equals the feed per revolution.

Figure 7.12b shows the same operation but using an *oblique* turning tool: that is, a turning tool in

which the cutting edge is inclined to the direction of feed. The hatched parallelogram represents the cross-sectional area df of the chip. When a tool is cutting obliquely the chip thickness is less than the feed per revolution.

For single-point cutting tools the area of the chip is *always* given by the expression:

$$\text{area of chip} = df$$

Figure 7.12c compares the cross-sectional areas for chips produced by orthogonal and oblique cutting:

(a) Although different in shape, all the chips have the same area. Thus the rate of metal removal is the same for any given spindle speed.

(b) With a tool cutting obliquely:
   (i) The chip is wider but thinner for a given area.
   (ii) Because the chip is thinner, it is more easily deflected over the rake face of the tool.
   (iii) This, in turn, reduces the cutting forces acting upon the tool and the frictional wear on the rake face of the tool.

(c) When a tool is cutting obliquely, there is a radial component of the feed force which tends to push the tool away from the workpiece. This keeps the flanks of the cross-slide screw and nut in constant contact, taking up any backlash and preventing the tool being drawn into the workpiece when taking heavy roughing cuts.

(d) Care must be taken when employing oblique cutting techniques. If the plan approach angle is too great then the chip will become so wide and thin that the tool will not bite into the work. Cutting will become intermittent, with chatter, and a poor surface finish will result.

Surfacing

Figure 7.13a shows the feed, depth of cut and cross-sectional area of the chip when turning with the direction of feed *perpendicular* to the workpiece axis, unlike the previous examples where the direction of feed was parallel to the workpiece axis. The cross-sectional area of the chip is still the feed per revolution f multiplied by the depth of cut d. In this example, oblique cutting is employed and the chip thickness is less than the feed. Again, this improves the cutting efficiency, prevents the tool being drawn into the work and ensures that a flat surface is produced. The cutting edge must be set exactly at the centre height of the work to ensure that the surface cleans up right to the centre.

Operator control of cut and feed

In operations such as parallel turning and surfacing

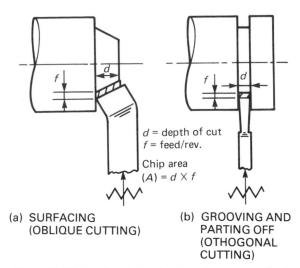

d = depth of cut
f = feed/rev.

Chip area
$(A) = d \times f$

(a) SURFACING (OBLIQUE CUTTING)

(b) GROOVING AND PARTING OFF (OTHOGONAL CUTTING)

Figure 7.13 Feed and depth of cut for perpendicular turning

on the lathe, the operator has control of both the rate of feed and the depth of cut. The selection of these variables is largely a matter of skill, experience and judgement, since the same chip area and rate of metal removal can be achieved by a deep cut and a fine feed or by a shallow cut and a coarse feed.

Deep cut, fine feed This lessens the number of passes required to reduce a component to a given diameter. It also leaves a good surface finish. The chip is being bent across its thinnest secion as it leaves the work and passes over the tool, thus reducing the load on the tool. Carried to extremes, too deep a cut with too fine a feed will cause difficulties for the cutting edge in penetrating the workpiece, resulting in intermittent cutting and chatter.

Shallow cut, coarse feed This enables the cutting edge to bite into the work easily, but a larger number of passes have to be taken (although these are taken more quickly) to reduce the workpiece to a given diameter. The finish will be rough. The chip will be thick; therefore the pressure of the chip on the rake face of the tool will be greater, causing more wear and making lubrication between the chip and the tool more difficult.

Thus, on balance, it is best to take a relatively deep cut at a moderate rate of feed.

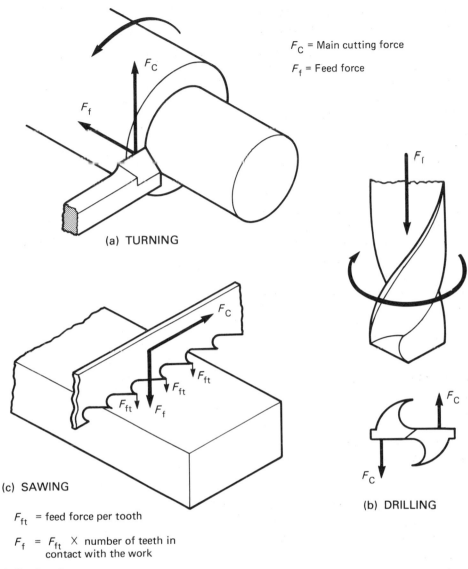

F_c = Main cutting force

F_f = Feed force

(a) TURNING

(b) DRILLING

(c) SAWING

F_{ft} = feed force per tooth

F_f = F_{ft} × number of teeth in contact with the work

Figure 7.14 Cutting forces

Grooving and parting off

In these operations the depth of cut is limited by the width of the tool. The depth of cut can never be less than the tool width, but it can be greater if roughing and finishing cuts are taken when grooving. Figure 7.13b shows examples of these operations.

The principles of definition of feed, depth of cut and cross-sectional area of the chip can be applied to other tools and operations, such as drilling, milling, shaping and sawing.

7.4 Forces acting on a cutting tool

During cutting operations, two main forces have to be exerted by the cutting tool in order to remove material. These are:

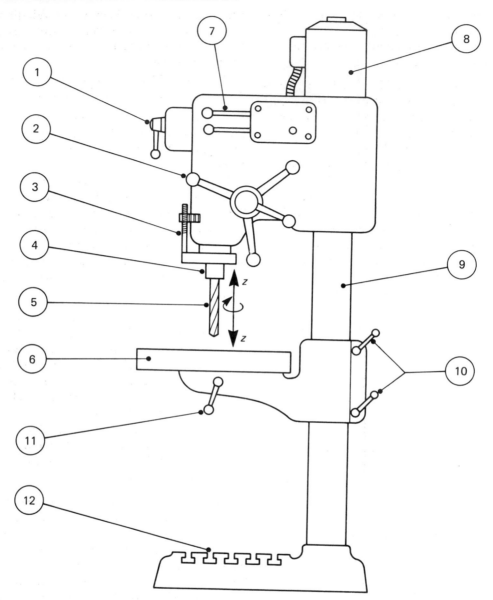

Parts of the Pillar Type Drilling Machine

1	Stop/start switch (electrics).	7	Speed change levers.
2	Hand or automatic feed lever.	8	Motor.
3	Drill depth stop.	9	Pillar.
4	Spindle.	10	Vertical table lock.
5	Drill.	11	Table lock.
6	Table.	12	Base.

Figure 7.15 Drilling machine

F_c The cutting force required to remove chips from the material being cut

F_f The feed force, which enables the cutting tool to penetrate the material being cut.

For simplicity, only tools with an orthogonal cutting action will be considered in this section, so that only these main forces have to be considered (there are additional cutting forces when tools cutting obliquely are used). The directions in which these forces act during turning, drilling and sawing are shown in Figure 7.14. It can be seen that where the tool has multiple cutting edges, the forces are shared. In the case of the twist drill, half the cutting force F_c is exerted on each lip of the drill (Figure 7.14b). In the case of the hacksaw, the forces exerted on each tooth are the main forces acting on the blade divided by the number of teeth. Conversely, the main forces acting on the blade are the forces acting on each tooth multiplied by the number of teeth. This is shown for the feed force in Figure 7.14c.

7.5 Drilling

Figure 7.15 shows a typical drilling machine. During a drilling operation the work is clamped to the table of the machine and does not move. The drill is held in the spindle of the machine, which rotates about the Z axis and also moves up and down along the Z axis. The downward or feed motion of the drill may be controlled by hand (sensitive feed) or by the machine mechanism (power feed). In addition to drilling holes, the drilling machine may be used for the following operations:

Countersinking A countersink bit is used to countersink a hole for rivets and screws. Countersink bits are usually supplied with an included angle of 90°.

Counterboring Counterboring tools are used to enable screws or bolts to lie below the surface of the work. The pilot acts as a guide and prevents deflection of the tool during cutting.

Spotfacing The spotfacing tool is similar to the counterboring tool. It is used to true up the surface of the work around a hole so that bolts or nuts can seat squarely in position.

Reaming A reamer is used to finish off a drilled hole to size. It has several cutting edges and produces a smooth accurate hole. It cannot correct axial misalignment.

Trepanning Trepanning cutters are mainly used for cutting large-diameter holes in sheet metal.

These operations are shown in Figure 7.16. In all the operations the work is fixed in position during cutting, and the only movement is that of the cutter about and along the Z axis.

The lathe may also be used for all the above operations as well as drilling. However, on the lathe, the work rotates about the Z axis and the drill merely feeds into the work along the Z axis. The drill does not rotate. The workpiece is held in the chuck of the lathe and the drill is held in the tailstock. There are serious disadvantages to this procedure:

(a) The taper bore of the tailstock is primarily to hold the tailstock centre and provide accurate location and axial alignment for the work to be turned. If the drill is allowed to grab and slip so that the tailstock bore becomes scored, then the fundamental alignment and accuracy of the lathe is affected.

(b) Only limited feed force is available, as the drill has to be fed into the work by hand by rotating the tailstock handwheel. There is no provision for applying power feed to the drill.

To ensure that the drilled hole is concentric and without wander, the hole is started using a centre drill held in a chuck. This provides a guide for the point of the twist drill. Because the feed force is limited, a pilot drill of smaller size should be put through first and then opened up gradually to size. Drilling is only a roughing-out operation and the hole should be finished by boring and/or reaming. Remember that a reamer only corrects the size and roundness of the hole. Only single-point boring can correct axial misalignment.

7.6 Parallel turning

The principal features of a centre lathe are shown in Figure 7.17

For turning parallel components the tool must

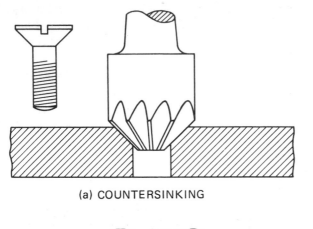

(a) COUNTERSINKING

(c) COUNTERBORING

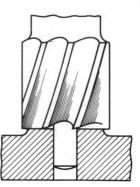

(d) SPOTFACING

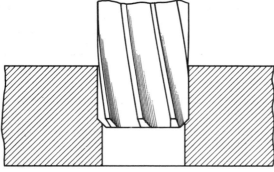

(b) REAMING

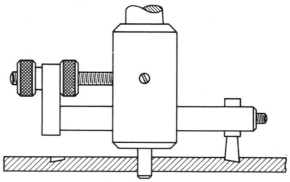

(e) TREPANNING

Figure 7.16 Further operations on the drilling machine

follow a path parallel to the principal axis of the lathe (the Z axis).

(a) The work is located and rotated about the Z axis by the spindle of the lathe. It may be held in a chuck, between centres, or on a face plate (see Chapter 6).

(b) The tool is moved parallel to the Z axis by the saddle or carriage of the lathe as it traverses along the bed slideways. This provides the feed of the tool into the work. Hand feed and power feed are available.

(c) The depth of cut is set by moving the tool along the X axis at right angles to the workpiece axis by means of the cross-slide. The depth of cut is set before cutting commences and remains constant throughout the turning

operation. These movements are shown in Figure 7.18.

When turning between centres, a trial cut should always be taken along the workpiece. If the headstock and tailstock centres are in alignment with the bed slideways, the work should be parallel and the diameter measured at the headstock end of the work should be the same as the diameter measured at the tailstock end of the work. If it is not, then the tailstock offset screw should be adjusted until the centres come into alignment and the work is parallel. Slender work or long work which has to protrude from the chuck should be further supported by the use of steadies, as shown in Figure 7.19.

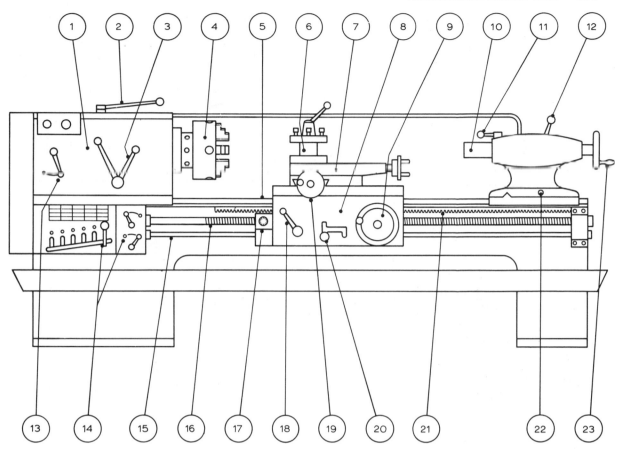

Parts of the Centre Lathe

1 All-geared headstock.
2 Clutch lever.
3 Speed change levers.
4 Chuck.
5 Bed slideways.
6 Four-way turret.
7 Compound slide.
8 Apron.
9 Sliding handwheel.
10 Tailstock sleeve.
11 Sleeve locking lever.
12 Tailstock clamping lever.

13 Feed reverse lever.
14 Quick-change gearbox.
15 Feedshaft.
16 Leadscrew.
17 Screwcutting dial.
18 Screwcutting lever.
19 Cross-slide handwheel.
20 Feed engage lever.
21 Rack.
22 Tailstock offset screw.
23 Tailstock handwheel.

Figure 7.17 Centre lathe

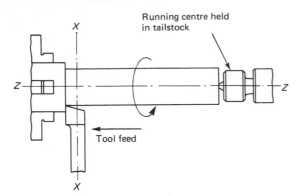

Figure 7.18 Parallel turning

(a) TURNING BETWEEN CENTRES

When turning cylindrical components on the lathe, it is usual for all the diameters to be turned concentric with each other: that is, all the diameters have a common axis. This is achieved by turning as many diameters at one setting as possible. If the work has to be removed and reset then it should be held in an independent-jaw (four-jaw) chuck and set to run true using a dial test indicator (DTI). Alternatively, if a batch of components is bring turned, the three-jaw self-centring chuck may be used with soft jaws previously bored out to run true.

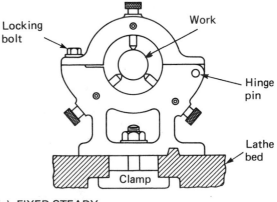

(b) TRAVELLING STEADY

7.7 Taper turning

Taper turning is the generation of conical surfaces. This is achieved by moving the tool along a path which is at an angle to the Z axis of the lathe. There are three techniques for doing this.

Compound slide

The compound slide (top slide) is mounted on the cross-slide and is capable of being rotated on a swivel base marked with a protractor scale. This enables the path of the tool to be inclined to the Z axis of the lathe, as shown in Figure 7.20a. This is a simple way of turning tapers and chamfers with a wide range of included angles. Unfortunately, use of the compound slide does have two disadvantages: no power feed is available, and the length of the taper is limited by the short travel of the slide. The depth of cut is set by the cross-slide.

(c) FIXED STEADY

Figure 7.19 Use of steadies

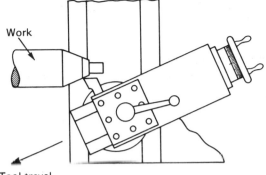

Work

Tool travel

(a) USING THE COMPOUND SLIDE

Guide bar   Slider

Taper turning attachment

Link   Bed

Work   Cross slide

Path of tool paralled
to guide bar which
is set to the half angle
of taper

(b) USING THE TAPER TURNING ATTACHMENT

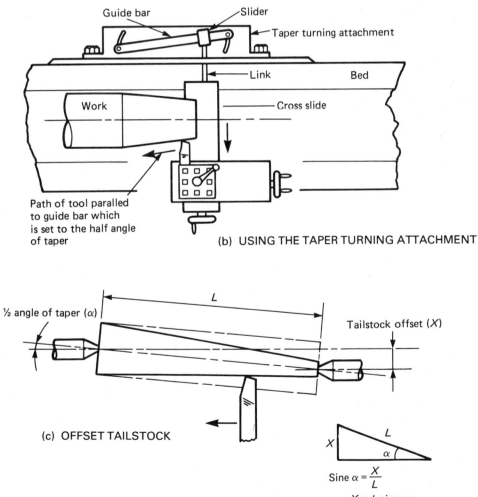

½ angle of taper (α)

L

Tailstock offset (X)

(c) OFFSET TAILSTOCK

X

L

α

$$\text{Sine } \alpha = \frac{X}{L}$$

$$X = L \text{ sine } \alpha$$

Figure 7.20 Taper turning

Taper turning attachment

This is an optional extra available for most centre lathes. An example is shown in Figure 7.20b. The guide bar on the back of the lathe is set over to the required angle. As the carriage moves along the bed of the lathe, a link from the slider on the guide bar of the taper turning attachment moves the cross-slide bodily to or from the work depending upon the direction in which it has been set. This arrangement has two advantages: the full length of the lathe may be used, and the normal power feed to the carriage is available. The disadvantage is the fact that only relatively slow tapers (small angles) may be generated. The depth of cut is set by the cross-slide in the usual manner.

Offset tailstock

It was stated in Section 7.6 that lack of parallelism in components turned between centres could be corrected by lateral adjustment of the tailstock. Similarly, tapered components can be produced by deliberately offsetting the tailstock as shown in Figure 7.20c. This technique has the advantage of enabling the full length of the machine to be used, and the normal power feed of the saddle is available. Unfortunately, only components which can be held between centres can be turned, and the angle of taper must be limited otherwise damage to the centres and centre holes will occur. Further this technique can only be applied to work supported on centres at each end. It cannot be used where the work is held in a chuck at one end. Ideally, spherically ended centres should be used.

7.8 Transverse turning: surfacing

When surfacing, the carriage is locked in position on the bed of the lathe. The feed to the cutting tool along the X axis is provided by the cross-slide, and the depth of cut is controlled by the compound slide, as shown in Figure 7.21. It is essential that the cutting tool is set at centre height so that there is no pip left at the centre of the workpiece. The work rotates about the Z axis in the normal way.

7.9 Transverse turning: grooving and parting off

When grooving and parting off, the compound slide only positions the tool. The depth of cut is

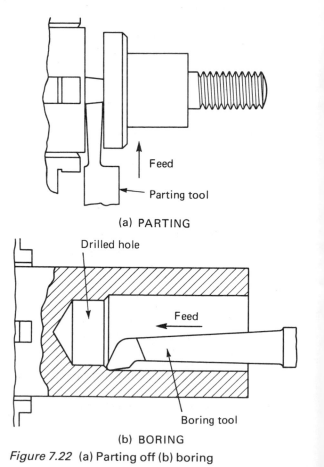

(a) PARTING

Drilled hole

(b) BORING

Figure 7.22 (a) Parting off (b) boring

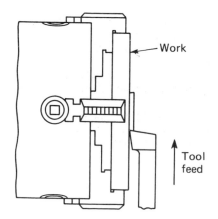

Figure 7.21 Surfacing

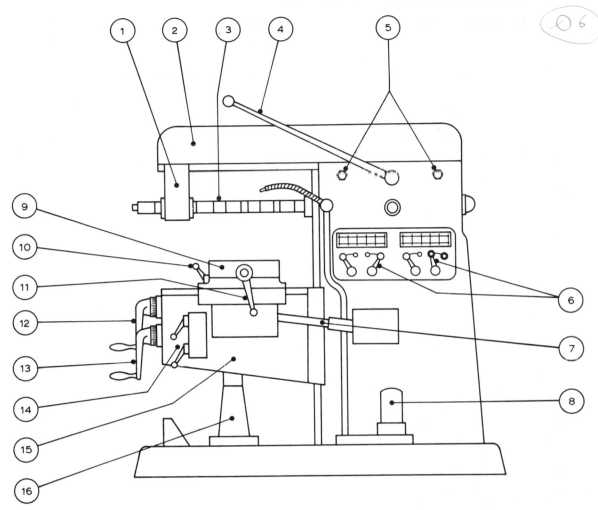

Parts of the Horizontal Milling Machine

1 Arbor support or yoke (which slides on overarm).
2 Overarm.
3 Arbor (on which the cutter is mounted).
4 Clutch lever.
5 Overarm clamping screws.
6 Speed and feed selector levers.
7 Telescopic feedshaft.
8 Coolant pump.
9 Table which contains tee slots for clamping work or a vice.
10 Automatic table feed (longitudinal).
11 Longitudinal feed (hand).
12 Vertical feed (hand) which raises or lowers knee assembly.
13 Cross or transverse handle (hand).
14 Vertical and horizontal automatic-feed levers.
15 Knee which supports table and moves up and down on dovetail slides.
16 Knee elevating screw.

Figure 7.23 Horizontal milling machine

controlled by the width of the tool (see Section 7.3). Parting off is a means of cutting off the workpiece from the bar stock (Figure 7.22a). Always part off as close to the chuck as possible for rigidity. For the same reason, keep the tool over-hang to a minimum. Tool clearances should be adequate but not excessive to avoid weakening the already slender tool and to avoid causing chatter. Lock the saddle whilst cutting, and maintain a steady feed. Do not allow the tool to rub.

7.10 Boring

This is used to enlarge drilled holes and to correct axial wander. Boring is a parallel turning oper-ation. The tools should be as rigid as possible to avoid deflection. However, they can never be as rigid as external tools, so care is necessary in selecting speeds, feed rates and depth of cut, particularly when deep holes are being bored. Ensure that there is sufficient secondary clearance to avoid the work being scored by the heel of the tool. Figure 7.22b shows an example of boring.

7.11 Horizontal milling machine

The essential features of a typical horizontal mill-ing machine are shown in Figure 7.23. This ma-chine generates flat surfaces, which may be parallel or perpendicular to the surface of the worktable, by means of rotating multi-tooth cutters. The relative movements of the work and the cutter along or about the X, Y and Z axes are shown in Figure 6.11c, and the methods of mount-ing the cutters are shown in Figure 6.14.

Figure 7.24 shows the basic principles of ma-terial removal using a milling cutter:

Conventional or upcut milling is the most common method of milling. Note the direction of rotation of the milling cutter. The cutting edge of the tooth can be seen cutting in an upward direc-tion. As the cutter is rotating in an anti-clockwise direction, the work is fed from the right against the cut. The direction of rotation will determine which side of the cutter the work should be fed into. Conventional milling tends to lift the work away from the vice or machine table. Always ensure that the work is secure.

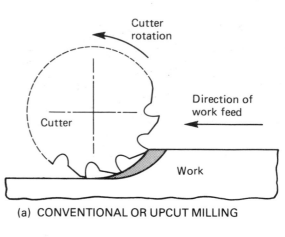

(a) CONVENTIONAL OR UPCUT MILLING

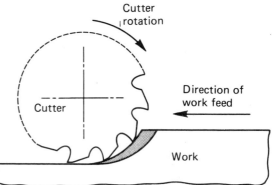

(b) CLIMB MILLING

Figure 7.24 Cutting action

Climb milling should never be attempted except under skilled supervision and on a machine specially equipped for this technique, as the cutter tends to climb over the work. The advantage of this method is that when taking heavy cuts the work is forced down on to the machine table. The feed force is also reduced.

Figure 7.25a shows a slab or roller milling cutter being used to produce a flat surface parallel to the worktable. Stepped and slotted components are produced by use of side and face milling cutters, as shown in Figure 7.25b. Slots are also produced by the use of slot milling cutters, as shown in Figure 7.25c. These do not have teeth on the side of the cutter, and wander less; however, they tend to clog in deep slots if too much material is removed for

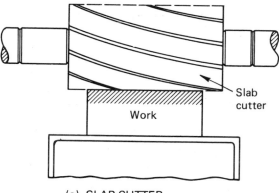

(a) SLAB CUTTER

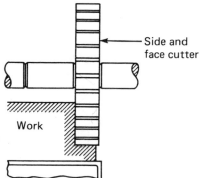

(b) SIDE AND FACE CUTTER

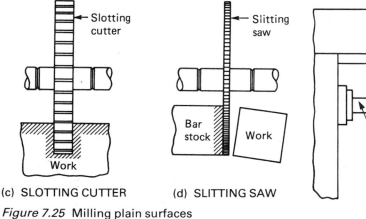

(c) SLOTTING CUTTER   (d) SLITTING SAW

Figure 7.25 Milling plain surfaces

each pass of the cutter. Narrow slots are cut with a slitting saw, as shown in Figure 7.25d.

Rigidity is of great importance for efficient machining. The difference between a bad and a good set-up is shown in Figure 7.26. Always pay particular attention to the position of the arbor support. This should be as close to the column of the machine as possible to avoid deflection of the cutter.

7.12 Vertical milling machine

The essential features of a typical vertical milling machine are shown in Figure 7.27. This machine

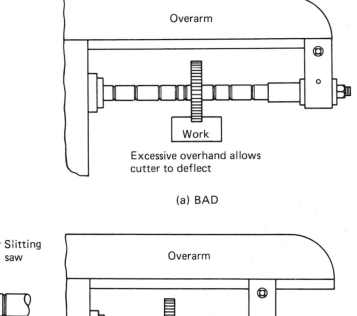

(a) BAD

(b) IMPROVED

Figure 7.26 Rigidity

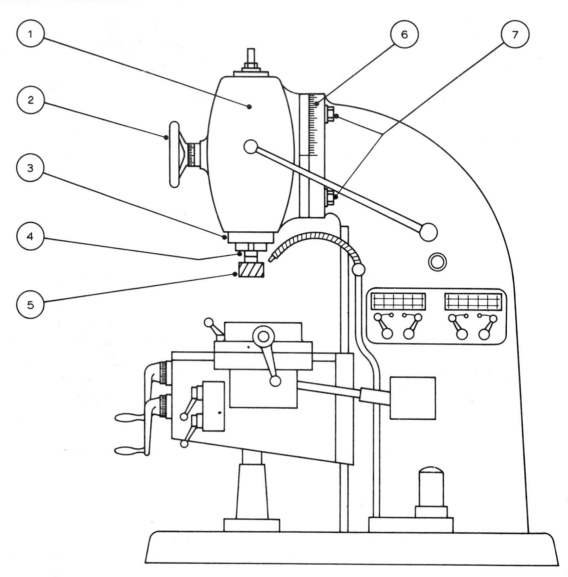

Parts of the Vertical Milling Machine

1 Vertical head which tilts.
2 Vertical feed handwheel.
3 Quill.
4 Spindle.

5 Cutter.
6 Head tilts here.
7 Head locking nuts.

Note: Remaining parts are similar to the horizontal-type milling machine.

Figure 7.27 Vertical milling machine: parts not numbered are similar to horizontal milling machine (Figure 7.23)

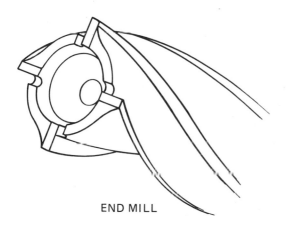

END MILL

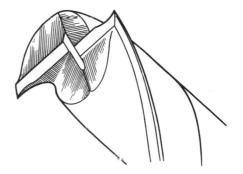

SLOT DRILL

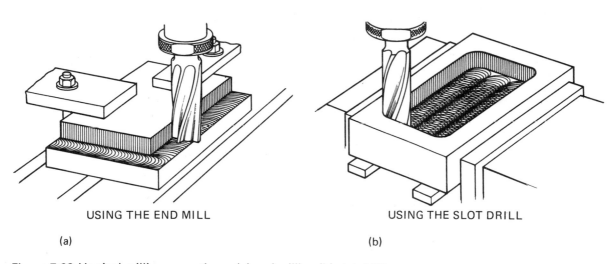

USING THE END MILL

(a)

USING THE SLOT DRILL

(b)

Figure 7.28 Vertical milling operations: (a) end milling (b) slot drilling

generates flat surfaces, which may be parallel, perpendicular or at an angle to the surface of the worktable, by means of rotating multitooth cutters. The relative movements of the work and cutter along or about the X, Y and Z axes are shown in Figure 6.11b, and the methods of mounting the cutters were shown in Figure 6.14.

Figure 6.14b shows a face milling cutter mounted directly into the spindle nose on a stub arbor in order to generate a surface parallel to the worktable of the machine. The lack of overhang results in a rigid mounting, and very much greater rates of metal removal are possible than with the slab mill on the horizontal machine. Figure 7.28a shows the formation of the cutting teeth on an end

mill, and a step being produced by such a cutter. Take care to feed the work at the correct side of the cutter. Vertical milling machine cutters are only supported at one end. Avoid climb milling on all but exceptionally light cuts. End mills are normally available with diameters from 3 mm to 40 mm; above this size, shell end mills are used. The formation of the teeth of end mills prevents them being plunged directly into the work; therefore they are unsuitable for pocket milling.

Figure 7.28b shows the formation of the cutting teeth of a slot drill, and a typical pocket milling operation using such a cutter. Slot drills are capable of being fed into the work like a twist drill, and can start a pocket milling operation from the solid.

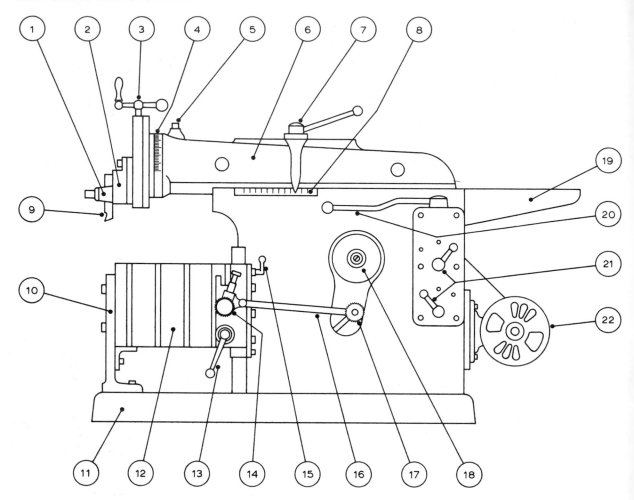

Parts of the Shaping Machine

1	Toolholder.	12	Table.
2	Clapper box.	13	Handle for vertical table movement.
3	Slide handwheel.	14	Horizontal feed ratchet.
4	Head swivels at this point.	15	Table clamp (vertical).
5	Locking screw for swivel head.	16	Feed connecting rod.
6	Ram.	17	Feed adjusting knob.
7	Ram clamping handle.	18	Stroke adjusting spindle.
8	Length of stroke indicator.	19	Ram guard.
9	Cutting tool.	20	Clutch lever.
10	Table support bracket.	21	Speed change levers.
11	Base.	22	Electric motor.

Figure 7.29 Shaping machine

7.13 Shaping machine

Figure 7.29 shows the main features of a shaping machine. It produces flat surfaces which may be parallel, perpendicular or at an angle to the surface of the worktable. The single-point cutting tool (which is similar to a lathe tool) takes a series of cuts parallel to each other and produces what is called a ruled surface, as shown in Figure 7.30. The tool is held in a toolpost which, in turn, is mounted on a clapper box as shown in Figure 7.31. The clapper box remains closed during the forward (cutting) stroke but lifts during the return stroke so that the tool can ride over the workpiece without damage. Figure 7.32 shows the setting of the clapper box for machining horizontal and vertical surfaces. Figure 7.33 shows further shaping operations, including angular surfaces.

When machining a horizontal surface, the feed is provided intermittently by incremental movements of the worktable during each return stroke of the ram and the tool. The worktable is stationary during the forward (cutting) stroke. The depth of cut is set by the handwheel and micrometer scale on the tool head slide.

When machining vertical and angular surfaces the intermittent feed to the worktable is stopped and it is adjusted by hand to control the depth of cut. Feed is provided manually by the handwheel of the tool slide. A few large machines have power feed available on the tool slide.

7.14 Sawing

Figure 7.34 shows a typical power hacksaw. The work is clamped in a vice and does not move during the cutting operation. The blade is held in the saw frame under tension. Downfeed is applied during each backstroke (cutting stroke), and the blade is lifted clear of the cut during the forward stroke to prevent the teeth rubbing. A coolant is flooded over the cut to cool and lubricate the blade in the slot and to flush away the chips to prevent clogging. A few small machines cut on the forward stroke. Care must be taken to verify which way a particular machine cuts so that the blade can be inserted with the teeth facing in the correct direction.

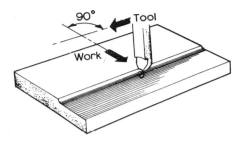
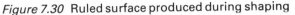

Figure 7.30 Ruled surface produced during shaping

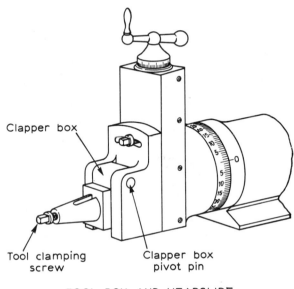

TOOL BOX AND HEADSLIDE

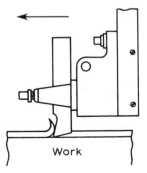
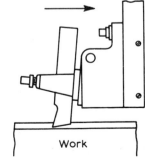

CUTTING STROKE   RETURN STROKE

Figure 7.31 Clapper box

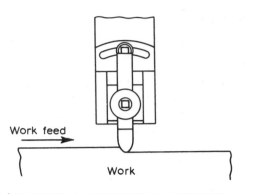

(a) SHAPING A HORIZONTAL SURFACE

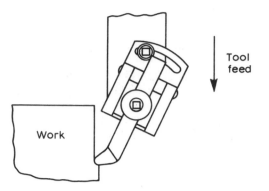

(b) SHAPING A VERTICAL SURFACE

Figure 7.32 Shaping horizontal and vertical surfaces

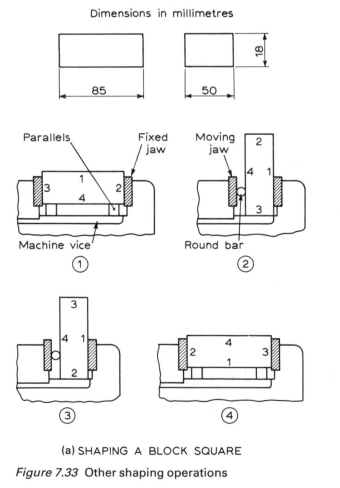

(a) SHAPING A BLOCK SQUARE

Figure 7.33 Other shaping operations

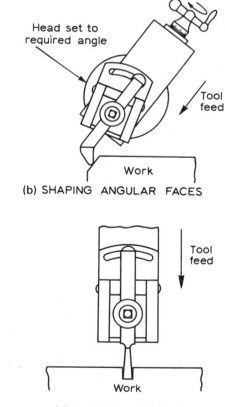

(b) SHAPING ANGULAR FACES

(c) SHAPING GROOVES

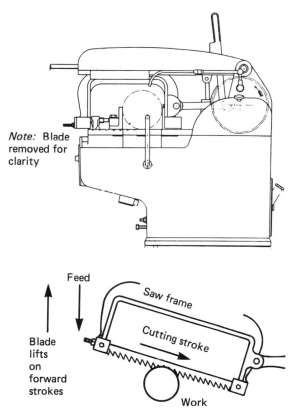

Note: Blade removed for clarity

Feed

Blade lifts on forward strokes

Figure 7.34 Power hacksaw

Exercises

For each exercise, select *one* of the four alternatives.

1 The correct names for the cutting angles of the single-point tool shown in Figure 7.35 are

(a) *A* = clearance angle; *B* = wedge angle; *C* = rake angle

(b) *A* = rake angle; *B* = wedge angle; *C* − clearance angle

(c) *A* = clearance angle; *B* = rake angle; *C* = wedge angle

(d) *A* = wedge angle; *B* = rake angle; *C* = clearance angle.

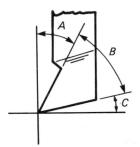

Figure 7.35

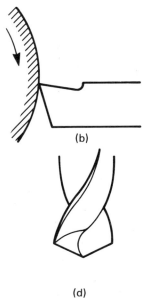

Figure 7.36

2 For a fixed clearance angle, increasing the rake angle of a cutting tool
  (a) increases the cutting efficiency but weakens the cutting edge
  (b) increases the cutting efficiency and strengthens the cutting edge
  (c) decreases the cutting efficiency and weakens the cutting edge
  (d) decreases the cutting efficiency but strengthens the cutting edge.

3 Which one of the cutting tools shown in Figure 7.36 has a *neutral* rake angle?

4 The angle of inclination for the chisel point shown in Figure 7.37 is
  (a) 21°
  (b) 24°
  (c) 31°
  (d) 55°

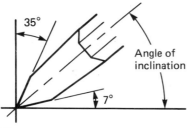

Figure 7.37

5 A new file should first be used on
  (a) soft materials such as brass before it is used on harder materials such as steel
  (b) harder material such as steel whilst it is still new and sharp
  (c) any material capable of being cut by a file
  (d) brittle materials before it is used on ductile materials.

6 When using a hacksaw to cut metal, tension the blade
  (a) using three turns of the wing nut, and cut with short quick strokes
  (b) by tightening the wing nut as far as possible, and cut using long steady strokes
  (c) by tightening the wing nut as far as possible, and cut using short quick strokes
  (d) using three turns of the wing nut, and cut with long steady strokes.

7 When a turning tool, cutting externally, is set above center, the effective
  (a) rake angle is increased and the effective clearance angle is decreased
  (b) rake and clearance angles are both increased
  (c) rake and clearance angles are both decreased
  (d) rake angle is decreased and the effective clearance angle is increased.

8 The heel of a boring tool is kept from rubbing on the surface of the bored hole without weakening the cutting edge by
  (a) raising the tool
  (b) increasing the clearance angle
  (c) using secondary clearance
  (d) tilting the tool.

9 The rake angle of a twist drill is controlled by
  (a) the helix angle
  (b) the point angle
  (c) the chisel angle
  (d) the land width.

10 The main purpose of the secondary clearance angle of a milling cutter tooth is to
  (a) provide chip clearance
  (b) prevent the heel of the tooth rubbing on the work
  (c) prevent the sides of the tooth rubbing when milling a slot
  (d) reduce the load on the tooth when milling a slot.

11 The turning tool shown in Figure 7.38 is said to be cutting
  (a) orthogonally
  (b) obtusely

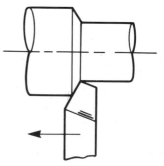

Figure 7.38

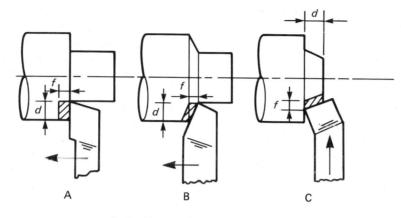

f = feed/rev   d = depth of cut

Figure 7.39

(c) obliquely
(d) acutely.

12 If the feed and depth of cut is constant in all the turning operations shown in Figure 7.39,
(a) the chip area at A is the greatest
(b) the chip area at B is the greatest
(c) the chip area at C is the greatest
(d) all the chip areas are the same.

13 The main forces acting on a turning tool when cutting orthogonally as shown in Figure 7.40 are
(a) 1 = feed force and 3 = cutting force
(b) 1 = cutting force and 3 = feed force
(c) 2 = feed force and 3 = cutting force
(d) 1 = cutting force and 2 = feed force.

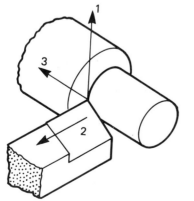

Figure 7.40

14 A hole is drilled slightly undersize and a reamer is then used to
(a) size and finish the hole
(b) correct axial misalignment (runout) of the drilled hole
(c) correct positional inaccuracies
(d) flat bottom the hole.

15 For normal drilling operations the spindle axis of a drilling machine should be
(a) perpendicular to the worktable
(b) inclined to the worktable
(c) parallel to the worktable
(d) rotating about the worktable.

16 When a turning tool moves parallel to the spindle axis of a lathe, the surface produced will be
(a) plane
(b) conical
(c) cylindrical
(d) tapered.

17 When producing short steep tapers on the centre lathe, it is usual to
(a) use the compound slide
(b) use the taper turning attachment
(c) offset the tailstock
(d) use the cross-slide.

18 In a shaping machine, the clapper box is provided to
(a) allow the tool to lift on the cutting stroke
(b) allow the tool to lift on the return stroke

(c) hold the tool rigid on both the cutting and return strokes

(d) allow the tool to lift on both the cutting and return strokes.

19 Milling machines are classified according to the
(a) plane in which the spindle axis lies
(b) plane of the table movement
(c) type of cutter which can be fitted
(d) types of arbor used.

20 To mill the pocket in the component shown in Figure 7.41 would require the use of
(a) an end mill
(b) a shell end mill
(c) a face mill
(d) a slot drill.

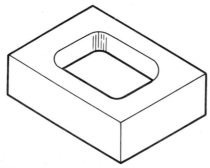

Figure 7.41

21 The milling technique shown in Figure 7.42 is known as
(a) downcut milling
(b) upcut (conventional) milling
(c) climb milling
(d) thread milling.

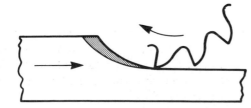

Figure 7.42

22 Which one of the settings shown in Figure 7.43 is correct for producing a vertical surface on a shaping machine?

23 To increase rigidity, most heavy-duty power hacksaws cut on the backstroke. Therefore the teeth of the blade
(a) should face forwards
(b) should face backwards
(c) can face in either direction
(d) require greater set.

24 On a power hacksaw, to remove the chips and prevent them clogging the teeth of the blade and to reduce wear on the blade generally,
(a) greater set is provided
(b) the rake angle of the teeth is increased
(c) compressed air is blown on the blade
(d) coolant is flooded over the cutting zone.

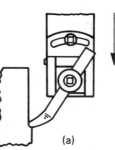

(a)

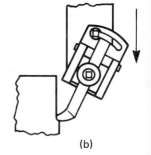

(b)

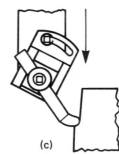

(c)

(d)

Figure 7.43

8

Joining

8.1 The purpose of joining

A motor car is an assembly of several thousand components, and many different joining techniques are required. Some of the components require replacement from time to time and are attached by temporary fastenings such as bolts and nuts. Others are permanent and are welded together. Some are semipermanent, such as the starter ring gear which is shrunk on to the flywheel. Increasingly, metals and non-metals are combined together in an assembly and these require special adhesive bonding techniques: for example, the flexible mountings for a car engine. Some components have to move relative to other parts of the assembly: for example, the doors on their hinges and the axles in their bearings. This chapter does not attempt to instruct the reader in the detailed procedures necessary to perform all the joining techniques which are used in engineering; it merely sets out to review the more important techniques and show some typical applications.

8.2 The range of joints

The joints used in engineering assemblies may be divided into the following categories:

Permanent joints A joint is considered to be permanent if one or more of the components, or the joining medium, has to be destroyed or damaged in order to dismantle the assembly.
Temporary joints A joint is considered to be temporary if the assembly can be dismantled with-

out damage to the components. It should be possible to reassemble the components using the original or new fastenings.
Flexible joints A joint is considered to be flexible if one component can move relative to another component in an assembly in a controlled manner: for example, a hinge.

8.3 Riveted joints

Riveting is a long-established method of making permanent joints using simple workshop equipment. Providing the joint is correctly designed so that the rivets are correctly proportioned and positioned and so that the rivets are stressed in shear and *not* in tension, then a very strong joint will result. Examples of typical riveted joints are shown in Figure 8.1. In addition to the selection of the most appropriate type of joint, it is also important to consider the following factors when selecting a rivet:

Material The material from which the rivet is made must not react with the components being joined to cause corrosion and weakening. This is particularly important when using light alloy rivets in airframe assemblies.
Choice of rivet head This is a compromise between strength and appearance.
Hole clearance If the clearance is too small, the rivet is difficult to insert and draw up; too large and the rivet buckles in the hole, resulting in a weak joint.

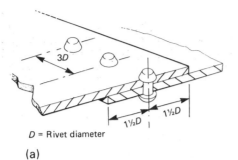

D = Rivet diameter

(a)

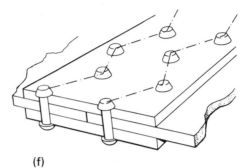

(e)

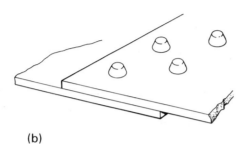

(b)

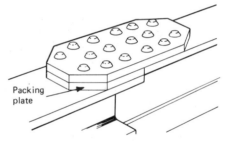

(f)

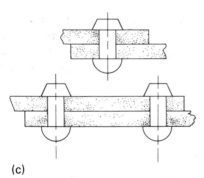

(c)

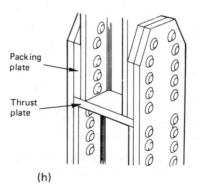

Packing plate

(g)

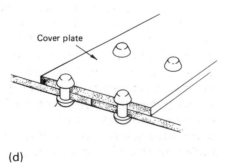

Cover plate

(d)

Packing plate

Thrust plate

(h)

Figure 8.1 Typical riveted joints: (a) single-riveted lap joint (b) double-riveted lap joint (c) assembly of lap joints (d) single-cover-plate butt joint (e) double-cover-plate butt joint (f) double-riveted, double-cover-plate butt joint, zigzag formation (g) splice joint (horizontal) (h) splice joint (vertical)

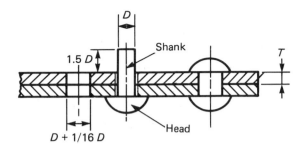

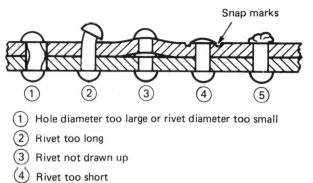

① Hole diameter too large or rivet diameter too small
② Rivet too long
③ Rivet not drawn up
④ Rivet too short
⑤ Head poorly shaped

Figure 8.3 Defects in riveted joints

Metal thickness (T)		Rivet diameter (ideal)
(mm)	(SWG)	(mm)
0.80	22	1.587
1.00	20	2.381
1.25	18	3.175
1.60	16	3.969
2.50	14	4.763
2.80	12	4.763 or 6.350
3.55	10	6.350 or 7.938

Figure 8.2 Proportions for riveted joints

Rivet length   If the rivet is too long, the rivet bends over during heading; too short and an inadequate head is formed. In both cases the result is a weak joint.

The proportions for rivet heads, hole clearance and rivet length are shown in Figure 8.2, whilst some typical defects in riveted joints are shown in Figure 8.3. The stages in closing a rivet are shown in Figure 8.4. It is not necessary to have the same type of head on each end of a rivet. All rivets should be heat treated to ensure maximum ductility before attempting to form the head, otherwise they will work harden and crack. Large rivets made from ferrous materials should be headed whilst red-hot: not only does this prevent cracking during the heading process but, as the rivet shrinks on cooling, the joint is pulled up tight.

For many applications, welded joints have superseded riveted joints. This is because:

(a) It is difficult to make riveted joints pressure tight.
(b) Riveting is slower than welding.
(c) Joint stress are uniformly distributed in welded joints, whereas they are concentrated at each rivet in a riveted joint.

(d) The components being joined are weakened at each point where a hole is drilled or punched to take a rivet.

Light-duty assemblies in sheet metal often use pop rivets. These are particularly useful for box sections where it is impossible to get inside to hold up the rivet during a conventional rivet heading process. Pop rivets can be installed from one side of the joint, as shown in Figure 8.5. They may be hollow for most applications, but solid rivets are available for fluid containers.

8.4 Compression joints

Mechanical

These joints rely upon the elasticity of metals to secure one component to another without the use of any additional fastening devices such as bolts, rivets or adhesives. In a lightly compressed joint, friction alone maintains assembly. In cases of extreme interference, one component may bite into the other with sufficient force for permanent deformation to take place, resulting in both friction and positive locking. The principles of pressed or staked joints are shown in Figure 8.6. In all compression joints it is essential that the outer component is made from a material which has a high tensile strength and high elasticity, whilst the inner should have high compressive strength. For example, it is in order to use steel for the outer component and cast iron for the inner. However, if

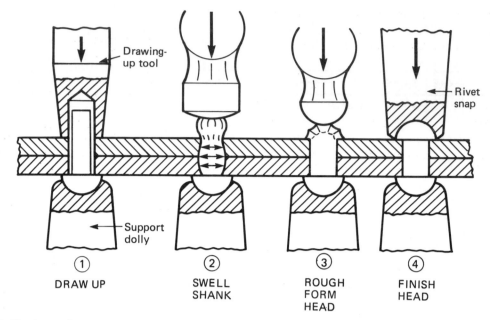

Figure 8.4 Closing a rivet

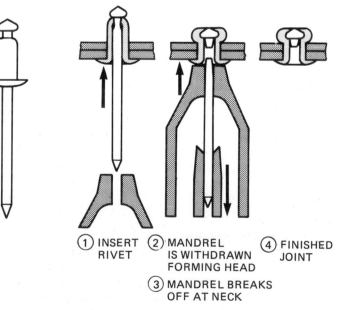

Figure 8.5 Pop riveting

these metals were reversed, the outer component would burst as the inner component was forced home because of the low tensile strength of cast iron.

Thermal

Compression joints may also be made by using thermal expansion and contraction instead of mechanical pressing to cause interference between the components to be joined. It is a property of metals that they expand when heated and shrink when cooled. This property can be exploited when making joints between two components. The advantages of this technique are that the joint is quickly made compared with such processes as brazing or welding, the properties of the materials are less affected, and dissimilar metals may be joined. Although the joint can be dismantled, it is usual to consider it a permanent joint since, invariably, one of the components has to be destroyed. For example, the starter ring gear of a motor car engine has to be split to release it from the flywheel. Because most metals are good conductors of heat it is virtually impossible to heat or cool one of the components alone once they have been assembled.

There are two types of thermal compression joint: hot shrunk and cold shrunk.

Hot shrunk joint

The principle of the hot shrunk joint is shown in Figure 8.7a. For example, this process could be used to fit an expensive phosphor bronze rim on to a low-cost cast iron or steel hub when making the wormwheel which is to mate with a steel wormgear. It would be too costly to make the wormwheel from solid phosphor bronze. To be successful the tolerances on the mating diameters of the two components must be closely controlled during manufacture, and heating must also be closely controlled and even. Too low a temperature will result in insufficient expansion, so that the rim will not slip over the hub. Too high a temperature can result in changes in the properties of the heated component.

Cold shrunk joint

The principle of cold shrinking is shown in Figure

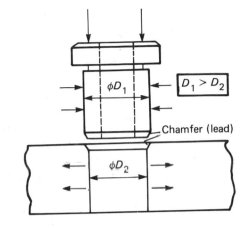

(a)

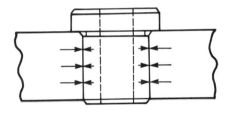

(b)

Figure 8.6 Press (compression) joint: (a) bush pressed home; outer expands and inner compresses (b) elasticity (spring-back) of assembled components results in very high friction forces, maintaining joint

8.7b. For example, this process could be used for fitting a starter ring gear of toughened alloy steel on to the flywheel of a motor car engine. To heat the ring gear sufficiently for it to fit over the flywheel could soften the gear and weaken the teeth. Thus it is better to cool and shrink the flywheel. Cooling does not permanently affect the properties of the metal, although it will become brittle whilst it is very cold. Its normal properties are restored when it returns to room temperature. Solid carbon dioxide or liquid nitrogen may be used as a coolant.

Care must be taken in handling heated or cooled components. The codes of practice must be observed, the tools provided for handling the components must be used, and any protective clothing provided must be worn.

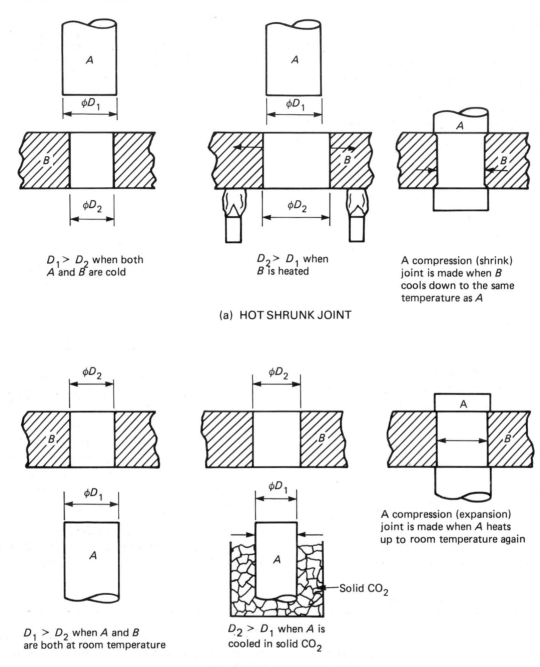

$D_1 > D_2$ when both
A and B are cold

$D_2 > D_1$ when
B is heated

A compression (shrink)
joint is made when B
cools down to the same
temperature as A

(a) HOT SHRUNK JOINT

$D_1 > D_2$ when A and B
are both at room temperature

$D_2 > D_1$ when A is
cooled in solid CO_2

Solid CO_2

A compression (expansion)
joint is made when A heats
up to room temperature again

(b) COLD EXPANSION JOINT

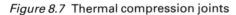

Figure 8.7 Thermal compression joints

8.5 Soft soldered joints

Soldering is the permanent joining of metals by the use of a filler metal of low melting point. In soft soldering this is an alloy of lead and tin and, depending upon the proportions of each, it melts at temperatures between 183°C and 288°C. In a correctly soldered joint a chemical reaction takes place between the solder and the parent metal of the components being joined to form intermetallic compounds at the interface. When this happens, the solder cannot be peeled from the joint face, nor can it be removed by melting. To form these compounds, which key the solder to the surfaces being joined, the surfaces must be clean and free from dirt, grease and surface oxides. A *flux* is used to produce this chemically clean surface. Figure 8.8a shows a soldered joint. Figure 8.8b shows the action of the flux, as follows:

A Flux solution lying above the oxidized metal surface
B Boiling flux solution removing the oxide film and etching the metal surface
C The bare metal in contact with the fused flux
D The liquid metal displacing the fused flux
E The tin in the solder alloy reacting with the parent metal to form a compound
F The solder solidifying.

There are two main types of flux:

Active (corrosive) Mineral acid base. These scour the joint surfaces thoroughly and give a sound joint easily. They leave a corrosive residue which has to be washed off the finished work; therefore they are unsuitable for fine work and the

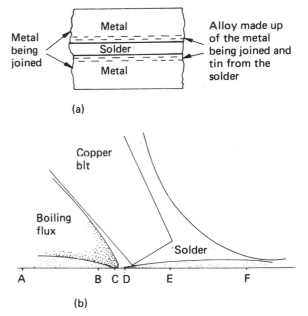

(a)

(b)

Figure 8.8 Principles of soft soldering: (a) section through a soldered joint (b) action of the flux

assembly of electrical and electronic components. An example is Baker's fluid, which is based upon an acidulated zinc chloride solution.

Passive (non-corrosive) Tallow, olive oil, resin. Resin in particular is used for fine electrical and electronic work. They do not prepare the joint surface as vigorously as the corrosive fluxes, and more care is required in the preparation of the joint surfaces when passive fluxes are used.

Some typical fluxes are listed in Table 8.1.

As soft solders have a low tensile strength, it is important that the joint should be closely fitting

Table 8.1 Some common fluxes for soft soldering

Flux	Type	Application
Killed spirit (zinc chloride)	Corrosive	The base of most proprietary fluxes; suitable for most metals
Ammonium chloride (sal ammoniac)	Corrosive	Useful for pretinning copper soldering bits
Dilute hydrochloric acid	Corrosive	Used on zinc galvanized materials
Phosphoric acid	Corrosive	Can be used effectively on stainless steels
Resin	Non-corrosive	Used extensively on electrical work
Olive oil	Non-corrosive	Used for soldering pewter
Tallow	Non-corrosive	Used with plumber's solder for wiping lead joints

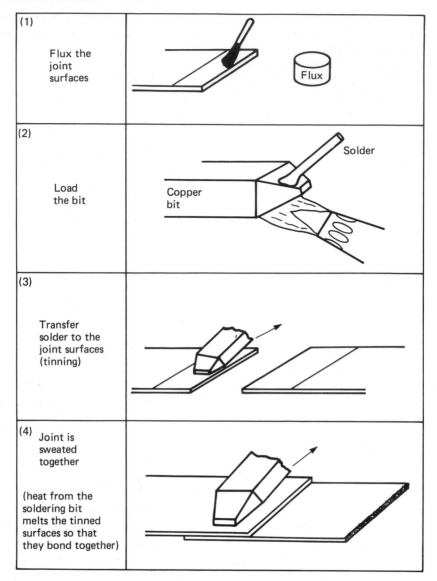

Figure 8.9 Producing a soft soldered joint

and the solder film kept to a minimum. A large area of contact is required. The stages in producing a soft soldered joint are shown in Figure 8.9.

8.6 Hard soldered joints

Hard soldering is a permanent joining process. The name 'hard soldering' is used for both the process of silver soldering and the process of brazing. Again, as in soft soldering, only the filler material or solder is melted. However, in hard soldering processes the filler material melts at much higher temperatures and the joint has a much greater strength, particularly at high temperatures. In a hard soldered joint the filler material is drawn into the joint by capillary attraction, so the clearance between the components must be closely controlled and the gap uniform. (Corner joints are not suitable for hard soldering.) As with soft soldering, preparation of the joint faces is of great importance. They should be scoured with a wire brush and an appropriate flux applied. The flux is usually based on borax mixed into a paste

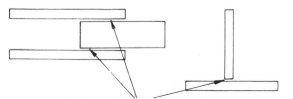

Gaps into which filler metal flows

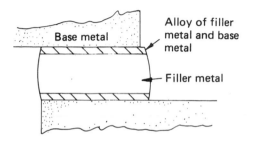

BONDING BY FORMING AN ALLOY AT JUNCTION

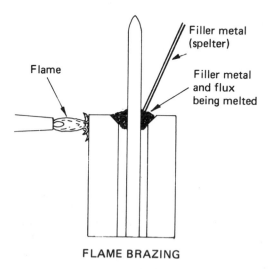

FLAME BRAZING

Figure 8.10 Principles of hard soldering

with water; proprietary filler materials and fluxes are available. Figure 8.10 shows the principles of hard soldering.

Silver soldering

This uses a copper-silver alloy with a melting temperature of 600 to 800°C depending upon the alloy used. A blowpipe is used to raise the temperature of the fluxed component to the required temperature. The solder should melt on contact with the component with the flame removed if the joint is at the correct temperature. The process can be applied to such metals as brass, copper, monel metal and stainless steel.

Brazing

This uses a copper-zinc (brass) alloy for the filler material: hence the name of the process. The filler material is called a spelter and, depending upon the composition of the alloy, has a melting point of about 870 to 980°C. Because of the high temperature involved, a gas torch (oxy-propane) is used as the heat source. For production brazing, controlled atmosphere furnaces and electric induction heating may be used. Again, a high-temperature flux such as borax is suitable. A properly brazed joint has a high tensile strength and is tough and ductile.

8.7 Fusion welded joints

These are permanent joints in which the components being joined are melted at the joint edges and additional molten filler metal is added. The filler metal is usually of similar composition to the metals being joined. Figure 8.11a shows the principles of fusion welding. High temperatures are involved and the heat-affected zone of the weld spreads into the components being joined, modifying the structure and properties of the metal.

The heat required to melt the base metal and the filler material can be supplied either by an oxy-acetylene flame or by an electric arc, as shown in Figure 8.11b. No flux is used for gas welding, and the molten metal is protected by the products of combustion of the flame which displace the air from the vicinity of the weld. In arc welding, a flux

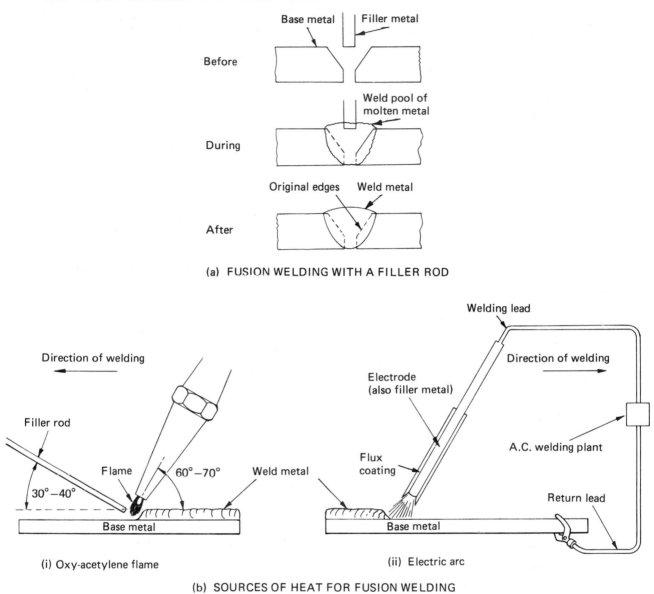

(a) FUSION WELDING WITH A FILLER ROD

(i) Oxy-acetylene flame

(ii) Electric arc

(b) SOURCES OF HEAT FOR FUSION WELDING

Figure 8.11 Principles of fusion welding

coating surrounds the electrode and helps to stabilize the arc as well as to protect the molten weld pool. During welding the flux and the electrode melt and the electrode provides the filler material. The hot flux gives off fumes, and adequate ventilation is required when arc welding.

Protective clothing must be worn when welding,

and goggles or a face mask with optical filters appropriate for the process must be used. The compressed gases used in welding are very dangerous, and welding equipment must only be used by trained and skilled personnel. Persons being trained must be continually and closely supervised.

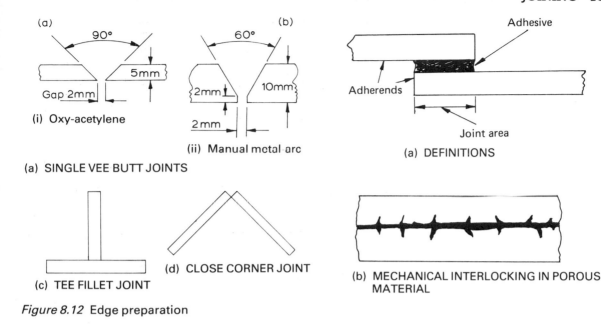

(a) SINGLE VEE BUTT JOINTS

(i) Oxy-acetylene

(ii) Manual metal arc

(c) TEE FILLET JOINT

(d) CLOSE CORNER JOINT

Figure 8.12 Edge preparation

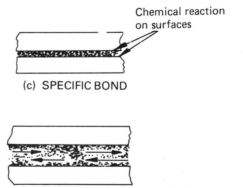

(a) DEFINITIONS

(b) MECHANICAL INTERLOCKING IN POROUS MATERIAL

(c) SPECIFIC BOND

(d) FORCES TRYING TO FRACTURE ADHESIVE

Figure 8.13 Principles of adhesive bonding

It is necessary that the edges of the parent metal are correctly prepared before welding commences. The purpose of this is to ensure good fusion of the parent metal with the filler metal and good penetration. This preparation may be done by shearing, flame cutting or machining, and the edges must be cleaned by degreasing, wire brushing or grinding. Figure 8.12 shows typical examples of edge preparation for welded joints.

8.8 Adhesive bonded joints

These are permanent joints in which an adhesive sticks (or adheres) two materials together. The bonding material is called the *adhesive*, the materials being stuck together are called the *adherends*, and the area of material adhered is called the *joint*. These definitions are illustrated in Figure 8.13a.

The advantages of adhesive bonding are:

(a) Similar or dissimilar materials may be joined.
(b) Joints are sealed and fluid tight.
(c) Adhesives are electrical insulators and inhibit electrolytic corrosion when dissimilar metals are joined.

(d) Temperature rise from the process is negligible and the properties of the materials being joined are unaffected.

The strength of an adhesive bond depends upon two factors:

Adhesion This can occur in two ways:
(a) Mechanical interlocking as shown in Figure

8.13b. This occurs when the adhesive penetrates the pores and surface irregularities of the adherends and produces a physical bonding or keying.

(b) Specific bonding as shown in figure 8.13c. This occurs when the adhesive reacts chemically with the adherends, bonding them through intermolecular attraction.

Cohesion This is the ability of the adhesive to withstand forces within itself, as shown in Figure 8.13d.

In addition it is essential for adhesives to have good wetting properties: that is, they should spread evenly over the joint face to form a continuous film and not ball up or run off the surface as shown in Figure 8.14.

Surface preparation is of great importance in adhesive bonding processes. The joint surfaces should be slightly roughened to improve adhesion. This can be done by the use of coarse emery cloth, wire wool, shot blasting, chemical etching and so on. Dust, scale and oxide films must also be removed from the joint surfaces. Any liquids present on the surfaces must be removed or they will prevent the adhesive adhering to the surfaces and penetrating into the pores of the surfaces. Grease and oil must also be removed by the use of solvents such as acetone, toluene or tri-chlorethylene. However, care must be taken in the use of such solvents as they are toxic and their vapours are highly flammable. They must only be used where adequate ventilation is provided and smoking is prohibited. Protective clothing must be worn as they are harmful to the skin. After the surface has been prepared it is important to ensure that it is not contaminated in any way, particularly by touching it with the hands.

For maximum joint strength it is important that the joint is designed so that it is only subjected to tensile or shear stress and that the load is evenly distributed. Figure 8.15a shows a joint in *tension*: all of the adhesive is employed in withstanding the tensile forces. Figure 8.15b shows a joint in *shear*: again, all of the adhesive is employed in withstanding the shearing forces. Figure 8.15c shows a joint in *cleavage*: this joint is not as strong as the previous two examples, as only the adhesive on the

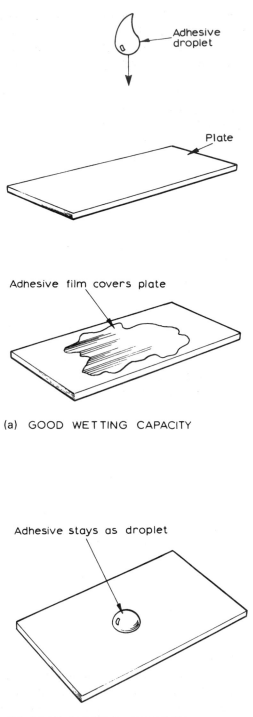

(a) GOOD WETTING CAPACITY

(b) POOR WETTING CAPACITY

Figure 8.14 Good and poor wetting capacity of an adhesive

right-hand side of the joint is holding the majority of the load. Figure 8.15d shows a joint in *peel*: this is the weakest of all the joints, only a very thin line of adhesive is withstanding the applied forces at any one time.

Figure 8.16 shows some typical joint designs for use with adhesives. Simple butt joints are not recommended, as the joint area is too small to withstand a reasonable load.

There are two basic groups of adhesives: thermoplastic and thermosetting.

Thermoplastic adhesives

These are based upon thermoplastic materials such as acrylics, cellulosics, rubber and vinyl products. These are materials which soften on heating and harden on cooling, the cycle being repetitive. They are the most widely used types because of their relatively low cost and ease of application. However, they are not suitable for highly stressed joints.

Thermoplastic adhesives consist of a thermoplastic material dissolved in a volatile solvent. They are applied in two ways:

(a) Direct application between the joint surfaces. Adhesion occurs when the volatile solvent evaporates. This can be unsatisfactory if the joint area is large, as the solvent at the centre of the joint may not evaporate.
(b) Impact adhesives. These are also thermoplastic materials dissolved in a volatile solvent. However, in this group the adhesive is spread on each joint face separately and allowed to dry. There is no hindrance to the evaporation of the solvent, no matter how large the joint area. When the solvent has evaporated, adhesion occurs by molecular attraction as soon as the joint faces are brought into contact.

Thermosetting adhesives

These are based on thermosetting plastics such as the phenolics, epoxies, polyesters, polyurethanes and silicones. They are hardened (polymerized) by curing. Curing requires the application of heat and pressure. Heat can be applied externally by an oven or an autoclave, or internally by the addition

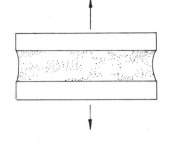

(a) JOINT IN TENSION

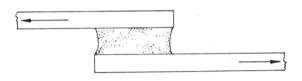

(b) JOINT IN SHEAR

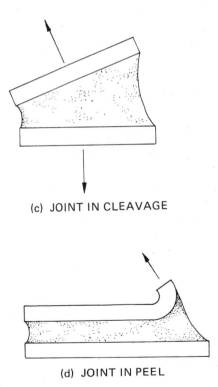

(c) JOINT IN CLEAVAGE

(d) JOINT IN PEEL

Figure 8.15 Strength of joints

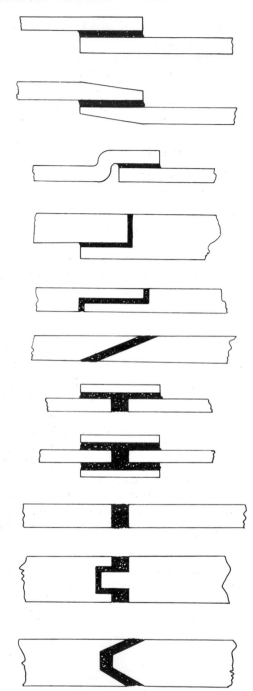

Simple lap

Bevelled lap

Rebated lap

Stepped lap

Double stepped lap

Scarf

Simple strap

Double strap

Simple butt

Tongue and groove butt

Scarf and groove butt

Figure 8.16 Joint design for adhesive bonding

of a chemical hardener. Once cured, these materials can never again be softened by the application of heat. Thermosetting adhesives are more expensive than thermoplastic adhesives, but they are very much stronger.

Table 8.2 indicates some adhesives suitable for bonding various materials. For example, if nylon is to be bonded to a structural metal, reference to the table shows that either a phenolic or an epoxy adhesive should be used.

8.9 Screwed fastenings

Screwed fastenings are the most widely used devices for temporary joints which have to be frequently assembled and dismantled. Figure 8.17a shows a hexagon head bolt and nut, and names its essential features.

Screwed fastenings are proportioned so that they require approximately the same force to make them fail in the following ways:

(a) By the head pulling off
(b) By the thread stripping (this assumes the nut is correctly proportioned)
(c) By failure in tension or shear across the minor diameter of the bolt thread.

The screw head converts rotary motion into linear motion, so that when the nut is tightened the bolt is slightly stretched. Providing the bolt is not stressed beyond its elastic limit, it will try to contract to its original length and thus will hold the joint faces tightly in contact. Figure 8.17b shows some typical screwed joints.

There are a number of factors to consider when using screwed fastenings, as described in the following sections. (Note: when replacing a worn or damaged fastening, one of identical specification to the original must be used.)

Material selection

Depending upon the application, the following properties must be considered:

(a) Adequate strength
(b) Corrosion resistance
(c) Compatability with the material of the components being joined, so that electrolytic corrosion of the components does not occur
(d) Electrical/thermal conductivity.

Screw thread system

There are a number of screw thread systems in existence:

ISO metric This is the international standard metric thread system, and it should be specified for all new designs. The proportions are shown in Figure 8.18. It is based upon a V-thread form with an included angle of 60°. A thread specified as M8 × 1 means that the screw will have a diameter of 8 mm and a thread of 1 mm pitch. The ISO metric thread system is available in coarse thread fastenings and fine thread fastenings. The importance of the pitch/diameter relationship is best

Table 8.2 Selection of adhesives

Adhesive type	Rubbers	Thermosetting plastic	PVC	Polyethylene (PTFE)	Polystyrene	Nylon	Cellulose	Acrylics	New structural metals	Structural metals
Thermoplastic										
Acrylics			√						√	
Epoxy based	√									
Cellulose based						√				
Rubber based	√	√	√			√	√	√		
Vinyl based	√	√	√			√	√	√		
Thermosetting										
Phenolic based	√	√	√	√		√				√
Epoxy based	√	√			√	√	√	√		√
Polyesters	√									
Polyurethanes										√
Silicones	√			√						

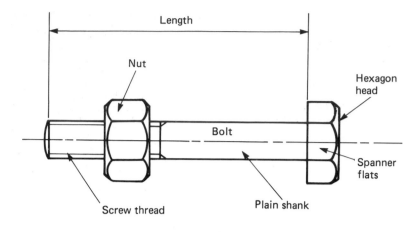

(a) HEXAGON HEAD BOLT AND NUT

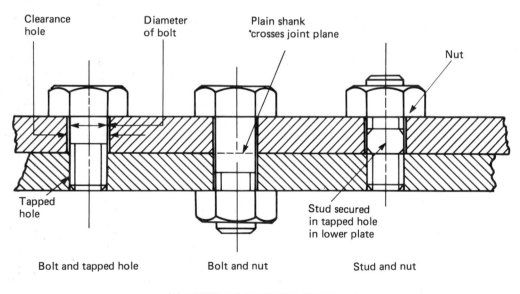

(b) TYPES OF SCREWED JOINT

Figure 8.17 Bolted joints

understood if it is accepted that the action of a screw thread is similar to the action of a wedge. The smaller the wedge action, the greater the pressure which can be applied by a given force. The wedge angles for two different pitches on the same diameter are shown in Figure 8.19. It can be seen that the fine pitch thread has a greater lifting or locking action, but the coarse pitch thread is stronger and provides more axial movement for each revolution. When threading aluminium

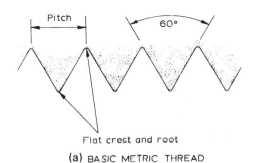

(a) BASIC METRIC THREAD

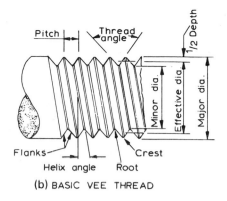

(b) BASIC VEE THREAD

Systeme International Screw Threads Series
(up to φ 30 mm only)

Major Dia.	Minor Dia.	Pitch	Depth	Flat	Effective Dia.	Tapping Size
6	4·70	1·0	0·65	0·12	5·35	5·00
(7)	5·70	1·0	0·65	0·12	6·35	6·00
8	6·38	1·25	0·81	0·16	7·19	6·60
(9)	7·38	1·25	0·81	0·16	8·19	7·70
10	8·05	1·50	0·97	0·19	9·03	8·40
(11)	9·05	1·50	0·97	0·19	10·03	9·50
12	9·73	1·75	1·14	0·22	10·86	10·25
(14)	11·40	2·00	1·30	0·25	12·70	12·00
16	13·40	2·00	1·30	0·25	14·70	13·75
(18)	14·75	2·50	1·62	0·31	16·38	15·25
20	16·75	2·50	1·62	0·31	18·38	17·25
(22)	18·75	2·50	1·62	0·31	20·38	19·25
24	20·10	3·00	1·95	0·37	22·05	21·00
(27)	23·10	3·00	1·95	0·37	25·05	24·00
30	25·45	3·50	2·27	0·44	27·73	26·50

Sizes in brackets are second choice (not preferred)

Figure 8.18 ISO metric (SI) screw threads

alloys, cast iron and plastic materials it is advisable to use coarse pitch threads where possible, as fine pitch threads are more easily stripped in such materials.

Unified (UNC and UNF) This thread system is widely used in the USA. It has the same propor-tions as the ISO metric system but is dimensioned in inch units.

British Standard Whitworth (BSW) This is an obsolete V-thread form with an included angle of 55°. Screwed fastenings with this thread form will be available as replacements for maintenance pur-

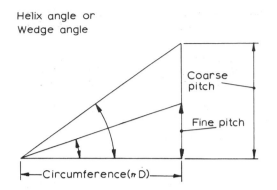

Figure 8.19 Pitch/diameter ratio

poses for some time to come, but are no longer used in new designs.

British Standard fine (BSF) This uses the Whitworth form but with a finer pitch. Fastenings with this thread are still available as replacements for maintenance purposes.

British Standard pipe (BSP) This has the original Whitworth form but is now given metric dimensions. It is available for fastenings with both parallel and taper threads.

British Association (BA) This system was originally introduced for miniature screwed fastenings for scientific instruments. It has a V-thread form with an included angle of 47.5°. This screw thread system has always been dimensioned in metric units, and is still widely used in electrical components. It is gradually being superseded by the smaller ISO metric sizes and the ISO miniature metric sizes.

Choice of head

There are a large variety of heads for screwed fastenings, and the selection is usually a compromise between strength, appearance and ease of tightening. The hexagon head is normally selected for general engineering applications. The cap screw is widely used for machine tools, jigs and fixtures, and other highly stressed applications. By recessing the head a flush surface is provided for safety and easy cleaning. The bolts on cap screws are flow formed from high-strength alloy steel and are very much stronger than hexagon bolts of the same size. Also the threads are more resistant to

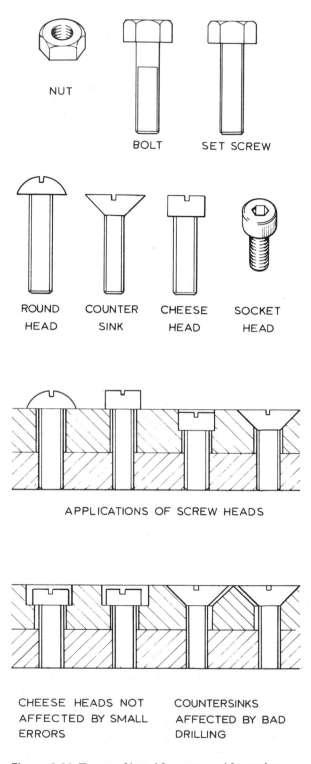

Figure 8.20 Types of head for screwed fastenings

wear during frequent dismantling and reassembly. Figure 8.20 shows a variety of heads.

Washers

Screwed fastenings must always pull down on to prepared seatings which are at right angles to the axis of the bolt, so that the bolt is not bent or distorted as it is tightened up. To protect the seating, a soft steel washer is inserted between the seating and the nut. This allows the nut to bed down uniformly and protects the seating from the scouring action of the nut as it is tightened. Taper washers should be used when erecting steel girders to prevent the draught angle of the flanges of the girders from bending the bolt.

Locking devices

These are used to prevent screwed fastenings from slackening off due to vibration. Such devices are essential in the control systems of machine tools, road vehicles, aircraft etc., where a joint failure would almost certainly result in loss of life. Locking devices may be *frictional* or *positive*.

Figure 8.21 shows a few of the many washers and locking devices available.

8.10 Pins, cotters and keys

These are all used to make temporary joints.

Dowels

Usually screwed fastenings are fitted through clearance holes, so unless fitted bolts are used to join two components, parallel dowels are required to provide positive location. Fitted bolts have ground shanks and are a light drive fit in the reamed hole. To provide this location, parallel pins or dowels are required. These are hardened and ground so as to be a light drive fit in a standard reamed hole.

Figure 8.22a shows a dowel fitted in position. Dowels should never be fitted in blind holes, not only because air is trapped under the dowel, making it difficult to fit, but also because it is impossible to drive out the dowel when the assembly needs to be dismantled. The dowel should have a taper lead so that, as it is driven home, it compresses slightly and the component

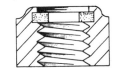
Fibre or plastic insert (frictional locking)

Slot for split pin (positive locking)

SELF-LOCKING NUT **CASTLE NUT**

PLAIN WASHER

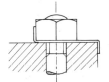
SPRING WASHER (FRICTIONAL LOCKING)

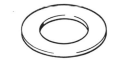
TAB WASHER **TAB WASHER IN USE (POSITIVE LOCKING)**

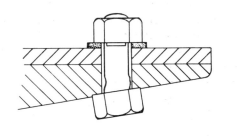

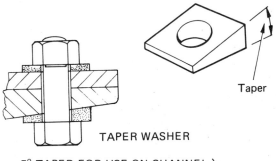
Taper

TAPER WASHER

5° TAPER FOR USE ON CHANNEL) FLANGES
8° TAPER FOR USE ON JOIST)

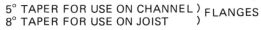

Figure 8.21 Washers and locking devices

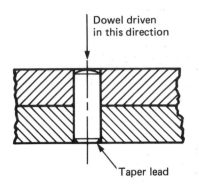

(a) PLAIN DOWEL IN PARALLEL REAMED HOLE

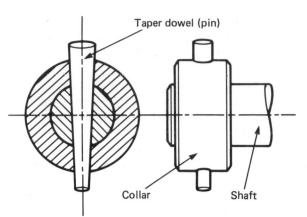

(b) TAPER DOWEL DRIVEN INTO A HOLE PREPARED WITH A TAPER PIN REAMER

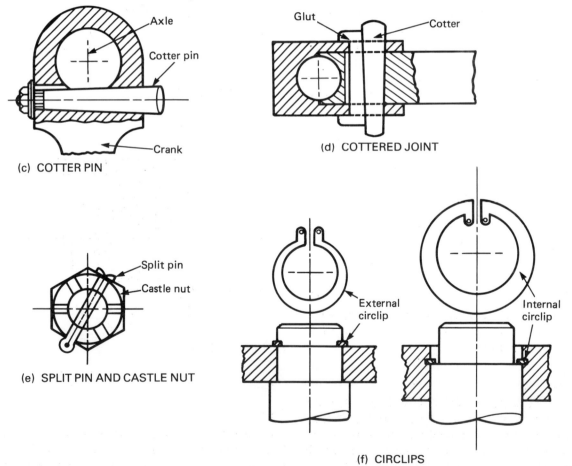

(c) COTTER PIN

(d) COTTERED JOINT

(e) SPLIT PIN AND CASTLE NUT

(f) CIRCLIPS

Figure 8.22 Miscellaneous fastenings

expands slightly such that the dowel is held in place by the elasticity of the materials (see mechanical compression joints in Section 8.4).

Taper pins

These are used for fitting components such as collars on to shafts, as shown in Figure 8.22b. When the collar is correctly positioned, a parallel hole is drilled through the collar and the shaft. This is opened up to the correct size by use of a taper pin reamer, and the taper pin is driven home. Any wear caused by dismantling and reassembly is compensated for by driving the pin in deeper. A Morse taper is used as on a twist drill shank. This is self-locking, so the pin will not drop out once it has been driven tightly home.

Cotter pins

These are taper pins secured by a nut, as shown in Figure 8.22c. The example shows the pedal crank of a bicycle fastened to its shaft by means of a cotter pin. Positive drive is provided by the flat on the pin. This flat is tapered so that wear can be taken up by tightening the nut and drawing the pin deeper into the joint.

Cotter

The cotter was once widely used as an alternative to screwed fastenings where a neat joint was required which could be quickly dismantled and reassembled. An example is shown in figure 8.22d. So that the taper cotter can be used in a parallel hole, a packing piece or glut is provided with the corresponding taper for the cotter to bed against.

Split pin

This is used for securing locking devices such as castle nuts and slotted nuts as shown in Figure 8.22e (see also Figure 8.21).

Circlips

These are used to provide a locating shoulder on a shaft or in a hole, as shown in Figure 8.22f. They are made from spring steel and are sprung into place by means of specially shaped pliers which locate in the holes of the circlip.

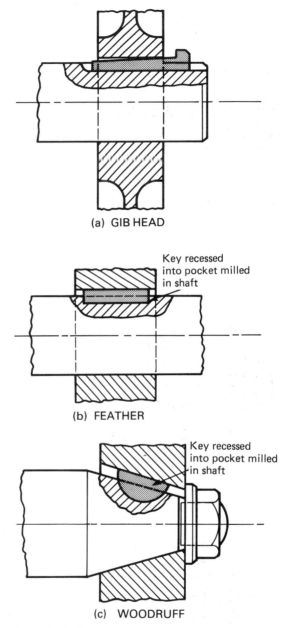

(a) GIB HEAD

Key recessed into pocket milled in shaft

(b) FEATHER

Key recessed into pocket milled in shaft

(c) WOODRUFF

Figure 8.23 Keys

Keys

These are used to connect wheels to shafts in order to transmit power. There are a number of different keys, and some of these are shown in Figure 8.23:

Gib-head (tapered) key This is driven into a slot cut half its depth in the shaft and half its depth in the wheel. Thus it can only be used where the wheel is fixed to the outer end of the shaft. Careful fitting is required, and the key and wheel are secured by friction although the power drive is positive. A taper drift can be used behind the head of the key to extract it.

Feather key This fits into a pocket milled in a shaft, and the key and the wheel may be positioned at any point along the shaft. The key is only fitted on its width and its clearance in depth. It only drives the wheel, and in no way secures it positionally.

Woodruff key This fits into a pocket which is a segment of a circle. This allows the key to float so that it is self-aligning. It is therefore suitable for use with taper seatings. A special milling cutter is required to make the pocket in the shaft.

8.11 Equipment and consumables

A number of the more common joining techniques used in engineering have been reviewed in this chapter. Before commencing to make the joint selected, it is essential to ensure that all the equipment and consumables required (such as welding gases and filler rods) are to hand and that the tools are in good condition and have been properly maintained. It is advisable to make a check list before commencing the job, particularly if it is on site and away from the stores.

Exercises

For each exercise, select *one* of the four alternatives.

1 Which of the following is a temporary joint:
   (a) A riveted joint
   (b) A welded joint
   (c) A shrink (compression) joint
   (d) A bolted joint?
2 Which of the following is a flexible joint:
   (a) A bolted and dowelled joint
   (b) A double-riveted lap joint
   (c) A hinged joint
   (d) A soldered joint?

3 Riveted joints are strongest when the rivets are stressed in
   (a) shear
   (b) tension
   (c) shear and tension combined
   (d) tension and compression combined.
4 The riveted joint shown in Figure 8.24 is known as a
   (a) lap joint
   (b) lap joint with cover plate
   (c) butt joint
   (d) butt joint with cover plate.

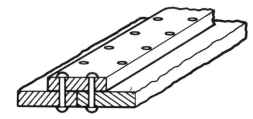

Figure 8.24

5 The ideal proportions for the rivet of length L and the hole diameter D shown in Figure 8.25 are
   (a) $L = 17.5$ mm and $D = 5.31$ mm
   (b) $L = 17.5$ mm and $D = 5.00$ mm
   (c) $L = 15.0$ mm and $D = 5.31$ mm
   (d) $L = 15.0$ mm and $D = 5.00$ mm.

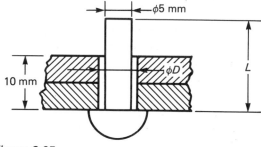

Figure 8.25

6 To rivet joints in closed box sections made from light-gauge sheet metal, it is most convenient to use
   (a) bifurcated rivets
   (b) pop rivets
   (c) tinmans rivets
   (d) snap head rivets.

7 When making compression joints with an outer
component and an inner component, which of
the following combinations of material properties are required:
   (a) outer with tensile strength, inner with ten-
       sile strength
   (b) outer with compressive strength, inner
       with compressive strength
   (c) outer with compressive strength, inner
       with tensile strength
   (d) outer with tensile strength, inner with
       compressive strength?

8 Drill bushes are mainly secured in the bush
plate of a drill jig by using a
   (a) compression joint (mechanical)
   (b) compression joint (hot shrunk)
   (c) compression joint (cold shrunk)
   (d) sweated joint.

9 Thermal compression joints are mainly depen-
dent upon the
   (a) expansion and contraction of metals when
       heated or cooled
   (b) thermal conductivity of metals
   (c) fusion point of metals
   (d) elasticity of metals.

10 Mechanical compression joints are mainly de-
pendent upon the
   (a) plasticity of metals
   (b) ductility of metals
   (c) thermal expansion of metals
   (d) elasticity of metals.

11 In a correctly soldered joint, the solder adheres
to the surface of the parent metal by
   (a) mechanical keying
   (b) the formation of intermetallic compounds
   (c) the formation of solid solutions
   (d) the adhesive effect of the flux.

12 Soldered joints in electronic equipment are
usually fluxed with
   (a) tallow
   (b) killed spirits
   (c) Baker's fluid
   (d) resin.

13 Sweated joints are made when
   (a) soft soldering
   (b) silver soldering
   (c) brazing
   (d) welding.

14 During hard soldering, the joint is at the
correct temperature when
   (a) the heat source melts the spelter or solder
   (b) the components being soldered are hot
       enough to melt the spelter or solder on
       contact
   (c) the edges of the joint start to melt
   (d) the combined heat of the heat source and
       the components being joined is required to
       melt the spelter or solder

15 A suitable flux for brazing is
   (a) resin
   (b) tallow
   (c) borax
   (d) zinc chloride.

16 The molten spelter is drawn into a brazed joint
by
   (a) the vacuum caused by contraction of the
       flux as it cools down
   (b) gravity
   (c) the lubricating action of the flux
   (d) capillary attraction.

17 When fusion welding
   (a) only the joint edges of the parent metal
       melt
   (b) only the filler metal melts
   (c) both the joint edges of the parent metal
       and the filler metal melt
   (d) neither the joint edges nor the filler metal
       melt.

18 When oxy-acetylene welding mild steel, no
flux is required because
   (a) the temperature of the joint is not high
       enough to cause oxidation
   (b) the products of combustion displace
       atmospheric oxygen from the vicinity of
       the weld pool
   (c) the filler material is self-fluxing
   (d) oxidation of the joint is unimportant when
       joining metals by fusion welding.

19 In arc welding, flux-coated electrodes are used
   (a) to stabilize the arc and protect the weld
       pool from oxidation
   (b) to economize on the amount of electricity
       used
   (c) to protect the weld pool from impurities in
       the filler metal

(d) to reduce heat loss and increase the temperature of the arc.

20 Joints for adhesive bonding are *weakest* when they are in
(a) tension
(b) shear
(c) cleavage
(d) peel.

21 Adhesion by specific bonding occurs only when
(a) one sort of material is being joined
(b) the joint relies upon mechanical keying of the adhesive and the adherends
(c) the adhesive reacts chemically ·with the adherends, resulting in intermolecular attraction
(d) metals are being bonded with adhesives.

22 The strongest adhesives are based upon
(a) natural glues
(b) thermosetting resins
(c) natural gums
(d) thermoplastic resins.

23 To hold two components in close contact, screwed joints rely upon the bolt having the property of
(a) high compressive strength
(b) ductility
(c) elasticity
(d) plasticity.

24 Torque wrenches are used to
(a) tighten up screwed joints as much as possible
(b) provide the fitter with increased leverage
(c) ensure that the fastenings take on a permanent set
(d) stress the screwed connection correctly.

25 Locking devices are used to prevent
(a) screwed fastenings working loose
(b) screwed fastenings being stolen
(c) the spanner slipping off whilst tightening the fastening
(d) the bolt dropping out if the nut works loose.

26 Screwed fastenings with fine threads are used where
(a) maximum thread strength is required
(b) maximum clamping and locking effect is required

(c) reduced production costs are required
(d) a low-strength material is being used.

27 Split pins are frequently used to secure
(a) collars to shafts
(b) castle nuts so that they will not work loose
(c) bearings to shafts
(d) lock washers in position.

28 Taper pins are frequently used to secure
(a) collars to shafts
(b) castle nuts so that they will not work loose
(c) bearings to shafts
(d) lock washers in position.

29 The most suitable and usual device for driving the pulley shown in Figure 8.26 is a
(a) gib-head key
(b) feather key
(c) cotter
(d) circlip.

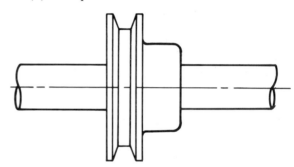

Figure 8.26

30 The self-aligning key shown in Figure 8.27 is a
(a) gib-head key
(b) feather key
(c) saddle key
(d) Woodruff key.

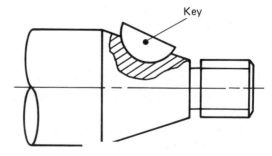

Key

Figure 8.27

9

Interpreting drawings, specifications and data

9.1 Purpose of technical drawings

Designers use technical drawings and specifications prepared by draughtspersons to convey their ideas and intentions to such people as:

Manufacturing engineers
Maintenance/service engineers
Sales engineers
Customers.

These technical drawings and specifications will vary depending upon for whom they are intended. The manufacturing engineer will want orthographic detail and assembly drawings. The service engineer will require exploded views. The customer will require installation drawings and operating data. Examples of these various types of drawing will be found later in this chapter.

Technical drawings and specifications should indicate such information as:

(a) The size and shape of the component
(b) The material from which it is to be made
(c) The finish (corrosion resistant, decorative etc.)
(d) The relationship between the component and associated components in an assembly.

In addition to the above, data must be provided to give guidance as to:

(e) The method of manufacture (operation sequence)
(f) The sequence for assembly and dismantling
(g) Installation and operation
(h) Performance compared with competing equipment for the benefit of the sales engineer.

To achieve all these requirements, various types of drawings are widely used. Visual representations avoid the confusion arising in translation from one language to another. Moreover, a drawing is more clear and concise and less open to misinterpretation than a verbal description. Most people have a natural instinct for appearance and proportion, and there is much truth in the saying that 'If a product looks right, it is right'. The designer must consider the function and fitness for purpose of a product; in addition, he or she must have a flair for styling and eye appeal, so that the product is attractive to the potential customer. This is particularly important when designing for the general public.

This chapter is concerned with the interpretation of technical information needed to manufacture, assemble and maintain engineering products, by the use of diagrams, drawings, charts and graphs, which together with the use of national and international standards and conventions provide a universal visual language.

9.2 Standards in technical communication

At the start of the Industrial Revolution there was no attempt at standardization. Each and every component was made individually. Each bolt and nut was made as a fitted pair and could not be interchanged with any other nut or bolt; this was highly inconvenient. Screw thread systems and the screwed fastenings using them were the first manufactured goods to be standardized, although only on a national basis at first; other standardized components followed. Since 1947 the International Organization for Standardization (ISO) has been steadily harmonizing national standards into fully interchangeable international standards, to ease and promote world trade in manufactured goods and services. (Measurement standards have been discussed in Chapter 4.) The British Standards Institution (BSI), as a major member of the ISO, is concerned with the standardization of goods and services in the UK and to ensure that those goods and services satisfy the requirements of world trade.

The BSI summarizes the main aims of standardization as:

(a) The provision of communication amongst all interested parties
(b) The promotion of economy in human effort, materials and energy in the production and exchange of goods
(c) The protection of consumer interests through adequate and consistent quality of goods and services
(d) The promotion of international trade by the removal of barriers caused by differences in national practices.

To achieve these aims the BSI publishes:

Standard specifications These specify the requirements for particular materials, products or processes, and they also specify the method for checking that the requirements of the standard have been met.

Codes of practice, guides etc. As their names imply, these are merely recommendations for good practice. They aim to assist in ensuring the provision of goods and services which are safe and of consistently high quality.

Glossaries To ensure that the standards are correctly interpreted, glossaries are available. These are lists of specialist and technical words and phrases together with their meanings. They are grouped according to the particular materials, products, processes and services to which they apply.

There are five categories of standardization which apply to the engineer and which assist in the understanding of technical drawings, specifications and published data. These are:

(a) Terminology and symbols (e.g. drawing conventions)
(b) Classification and designation
(c) Specifications for materials, standard products, and processes
(d) Methods of measuring, testing, sampling and analysing
(e) Recommendations on product or process applications and codes of practice.

ISO and BSI standard specifications are the result of consultation and agreement between the interested parties. They only become legally binding if they form part of a contract, or if a claim of compliance is made, or if they are included in legislation.

9.3 Methods of communicating technical information

Communications should always be: easy to understand and not subject to misinterpretation; sufficiently detailed that no additional information needs to be sought; but without repetition and without redundant information which might prove confusing.

Technical drawings

These may range from preliminary freehand sketches to formal drawings produced to BS 308: 'Engineering drawing practices'. Since technical drawings have to represent three-dimensional solids on a two-dimensional sheet of paper, various techniques have to be used. These include pictorial views such as isometric and oblique, and

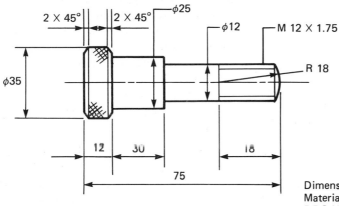

Dimensions in millimetres
Material: Free-cutting mild steel
DRG No: 4−950−3

A. N. ENGINEERING					
DRAWING NUMBER	BATCH	MATERIAL	MATERIAL SIZE	DATE REQUIRED	AUTHORISED
4-950-3	5	Free cutting mild steel	⌀ 35mm bar	30/6/1988	RLT
REMARKS					
Finish – self colour. Remove all burrs. Clean and oil.					

OP. No.	DESCRIPTION OF OPERATION	SPEED rev/min	FEED mm/rev	TOOLING
	Use Centre-lathe.			
1	Chuck bar	—	—	3-jaw chuck
2	Face end	320	hand	facing tool
3	Turn 25 mm diameter	400	0.05/rev	bar turning tool
4	Turn 12 mm diameter	810	0.05/rev	bar turning tool
5	Radius end	320	hand	form tool
6	Thread	45	hand	die-holder in tailstock
7	Chamfer	320	hand	facing tool
8	Knurl 35 mm diameter	45	hand	knurling tool
9	Part off	240	hand	parting tool
10	Reverse in chuck and hold on 25mm diameter	—	—	3-jaw chuck
11	Face to length	320	hand	facing tool
12	Chamfer	320	hand	facing tool

Figure 9.1 Operation sheet

orthographic drawing where true shapes and sizes are required.

Operation sheets

Figure 9.1 shows a typical operation sheet for a simple component. The purpose of an operation sheet is to indicate to the person making the component the tasks which need to be done, the sequence in which they are done, and the tools required to do them. In this example it also specifies the speeds and feed rates to be used.

Data sheets, wall charts and tables

The design engineer, the technician and the craftsperson have to refer to technical data frequently in their work, and this information needs to be readily to hand. Many manufacturers publish data sheets and wall charts concerning the selection and use of their products. These are not only very useful but are a powerful advertising medium. A typical chart may give feeds and speeds for different drill sizes when drilling various common materials.

British Standard specifications are issued in booklets. Electrical installation engineers will constantly refer to the 'Regulations for the electrical equipment of buildings' published by the Institution of Electrical Engineers (the IEE Regulations). Many tables of data are collected together and published in book form.

Graphs

Line graphs Just as engineering drawings are used as a clear and convenient way of describing complex products, so can graphs be used to give a clear and convenient description of numerical and mathematical data. Figure 9.2a shows a graph of the relationship between distance s and time t for the mathematical expression $s = at^2/2$ when the acceleration a is 10 m/s². In this instance it is correct to use a continuous flowing curve connecting the points plotted, since not only do these points lie on the curve, but every corresponding value of t and s between the points plotted also lies on the curve. However, this is not true for every type of graphical representation. Figure 9.2b shows a graph relating velocity and time. From A

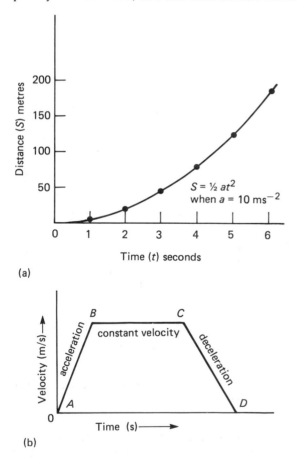

(a)

(b)

Figure 9.2 (a) Graph of a mathematical relationship (b) graph connecting three events which are mathematically unrelated

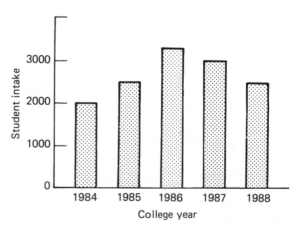

Figure 9.3 Histogram

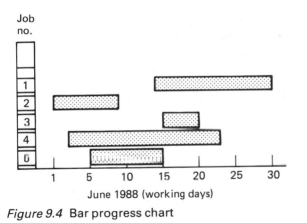

Figure 9.4 Bar progress chart

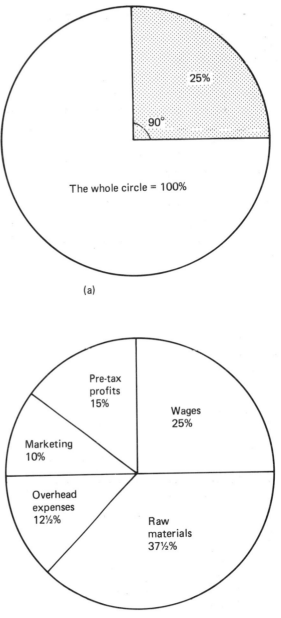

(a)

to B the object is accelerating; from B to C it is travelling with constant velocity; and from C to D it is decelerating (slowing down). In this example it is correct to connect the points with straight lines, since each stage of the journey is represented by a linear expression which is unrelated to the preceding or to the following stage of the journey.

Histograms Consider the student intake of a college over a number of years. The students enrol in September and leave in July. Between these months there is negligible change in the number of students attending. Therefore to plot the enrolments for each September and connect these points by a flowing curve of the type shown in Figure 9.2a would be incorrect, since it would imply that there is a continuous change in student population between one September and the next, and that this change satisfies a mathematical equation. It would also be incorrect to connect the plotted points by straight lines of the type shown in Figure 9.2b. This would imply that although the

Number of cars using a carpark each week

Figure 9.5 Ideograph (pictogram): number of cars using a car park each week

(b)

Figure 9.6 Pie charts: (a) principle (b) expenditure of a company

points are not following a mathematical expression, nevertheless there is some continuous change in the number of students enrolled between one September and the next. The correct way to plot this type of information is the histogram shown in

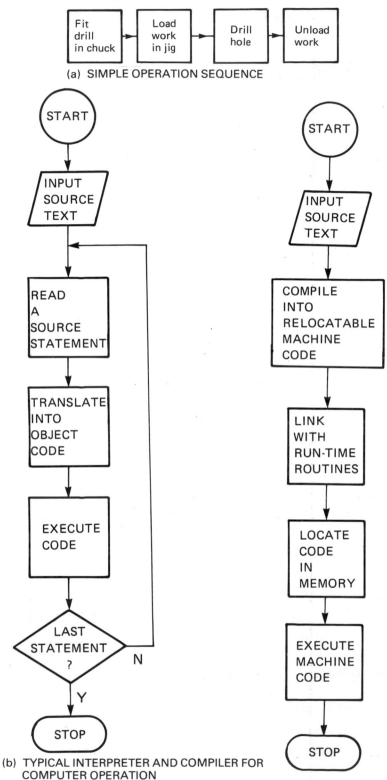

(a) SIMPLE OPERATION SEQUENCE

(b) TYPICAL INTERPRETER AND COMPILER FOR COMPUTER OPERATION

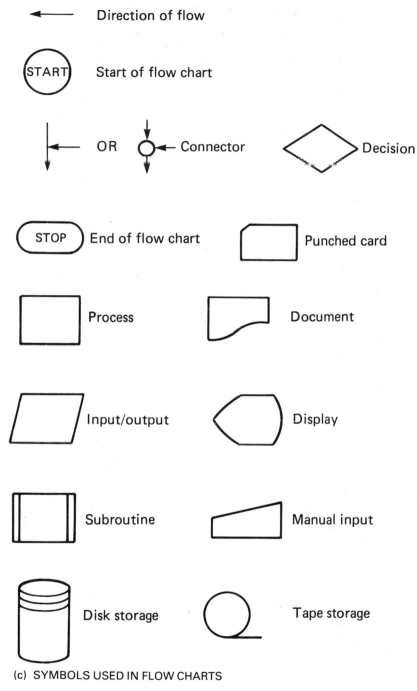

(c) SYMBOLS USED IN FLOW CHARTS

Figure 9.7 Flow charts

Figure 9.3. This clearly compares the enrolment trends over a number of years, but indicates that the enrolments for each year are unrelated by any mathematical equation.

Bar charts are also used for statistical information, but are usually plotted horizontally as shown in Figure 9.4.

Ideographs (pictograms) are frequently used for presenting statistical data to the general public, as shown in Figure 9.5. In this example each symbol represents 1000 cars. Therefore, in 1985, 3000 cars used the car park each week (three symbols each representing 1000 cars).

Pie charts are used to show how a total quantity is divided into its individual parts. For example, since a complete circle is 360°, 25 per cent of the total is $360 \times 25/100 = 90°$, as shown in Figure 9.6a. Figure 9.6b shows how the expenditure of a company can be represented by a pie chart.

Flow charts are used to illustrate a sequence of events. Figure 9.7a shows a very simple flow chart for drilling a hole in a component. Figure 9.7b shows two examples of flow charts as used in computer programming. The shapes of the boxes containing the instructions have particular meanings: these are explained in Figure 9.7c.

Data storage

Improvements in photographic techniques over the last few years have resulted in important developments in data storage and retrieval systems. Engineering drawings can be stored on *microfilm*. The large drawings are reduced photographically on to 16 or 35 mm film for storage, and are projection printed in full size when required. This saves considerable storage space in a drawing office.

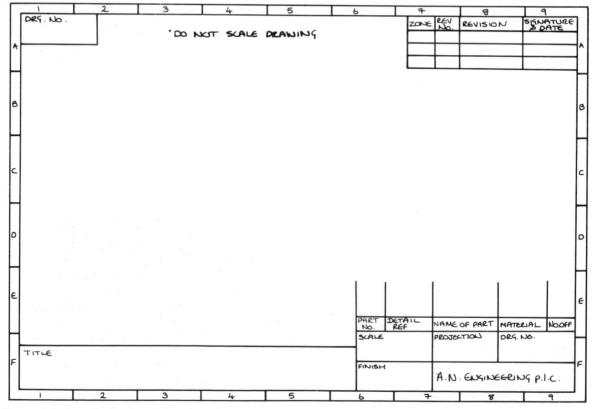

Figure 9.8 Standard printed drawing sheet

Libraries and stores use *microfiche* systems. Data is stored photographically in a grid of frames on a large rectangle of film. A desktop viewer is used to select and enlarge a single frame and project the data on to a rear projection screen for easy reading.

Electronic systems are also being used increasingly with computer terminals, to provide not only technical data of goods in store, but also an up-to-date position of the stock available. Video-tapes are also widely used for sales campaigns, with the sound track dubbed in different languages. It is easier and cheaper to ship a videotape to a customer abroad than to take a large machine. The use of videotapes depends upon the international standardization of video recording systems – another example of the importance of standardization.

9.4 Interpretation of technical drawings

Figure 9.8 shows the layout of a typical engineering drawing sheet. To save time these are usually printed to a standardized layout for a particular company, ready for the draughtsperson to add the drawing and complete the boxes and tables. As well as the drawing, the following additional information is also found on engineering drawings:

The projection (first or third angle)
The unit of measurement (inch or millimetre)
The scale
The material and specification
Heat and protective treatments
General tolerance
Tool references
Warning notes.

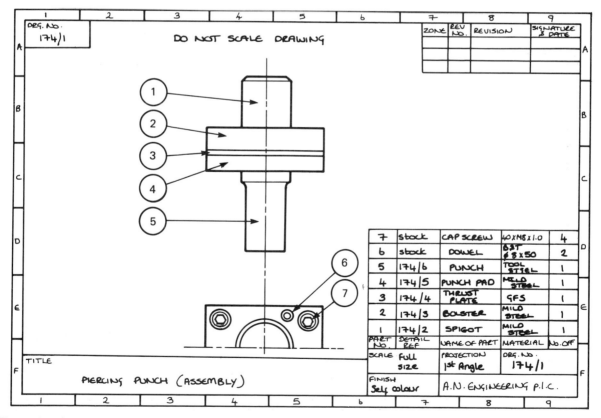

Figure 9.9 Assembly drawing

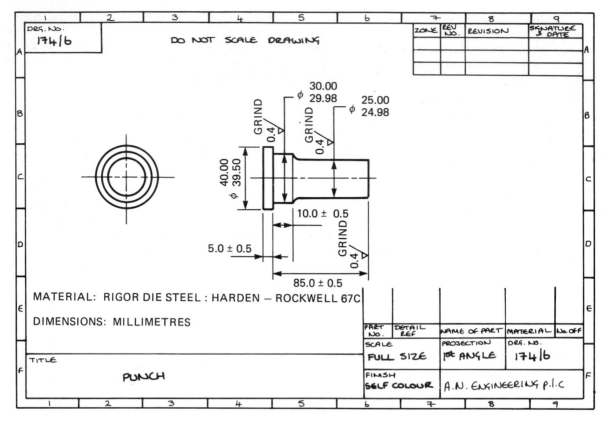

Figure 9.10 Detail drawing

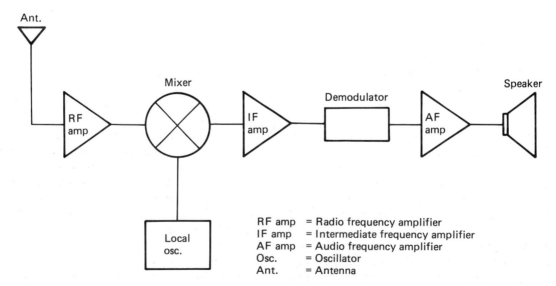

Figure 9.11 Block plan

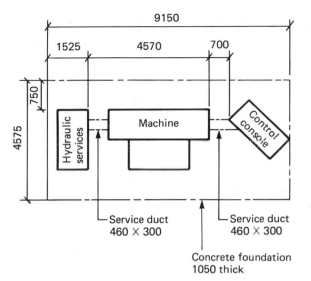

Figure 9.12 Location drawing

There are various types of engineering drawing depending upon the use for which they are required.

Assembly drawing This shows all the components correctly assembled together. Dimensions are not usually given on assembly drawings. All the parts are listed together with the quantity required. Manufacturers' catalogue references may also be included for bought-in components. The detail drawing reference numbers are also included where these have to be manufactured. An example is shown in Figure 9.9.

Detail drawing As the name implies, a detail drawing provides all the details required to manufacture a component. The detail drawing for the presstool punch is shown in Figure 9.10. It can be seen that it shows the shape and toleranced dimensions for the component, the material from which it is to be made, any subsequent heat treatment and decorative or corrosion-resistant finish, and any other essential information. Additional detail drawings would also be required for the punch pad, thrust plate, bolster plate and spigot.

Block plans These are used to show the main subassemblies and their individual relationship. An example for a radio receiver is shown in Figure 9.11.

Location drawings These are used to show where the various items of a major installation are located in relationship one to another. It may also include the foundation plans. A simple example is shown in Figure 9.12.

9.5 Orthographic projection

Engineering drawings are usually drawn in either first-angle or third-angle orthographic projection, as shown in Figure 9.13. This is a technique for representing three-dimensional solid objects on a two-dimensional sheet of paper so that all the views are of true shape and true size. Figure 9.13a shows a drawing projected in the first angle, whilst Figure 9.13b shows the same object projected in the third angle. Orthographic projection is used for detail drawings and assembly drawings where they will be read by engineers skilled in their interpretation.

Section views

The hidden detail in hollow components can often be shown more clearly by use of a section view. Figure 9.14 shows how drawings are sectioned. Imagine that the cutting plane actually cuts through the component along a specified plane. The unwanted portion of the component is removed and the remaining portion shows the sectioned view. Note that cut surfaces are sectioned shaded (hatched) to indicate where the component has been cut.

9.6 Conventions

Standard conventions are used to avoid the detailed drawing of common features in frequent use. These conventions are a form of drawing shorthand. They may be to company standards, national standards and international standards. Those generally used in the UK are the conventions specified in BS 308: 'Engineering drawing practice'. These, in turn, are harmonized with ISO standards, and are therefore international as well as national.

Examples of standard lines and their uses are shown in Figure 9.15. Examples of conventions for common features are shown in Figure 9.16.

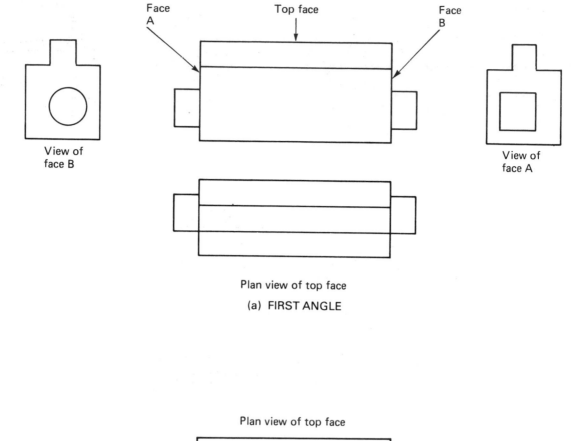

View of
face B

Face
A

Top face

Face
B

View of
face A

Plan view of top face

(a) FIRST ANGLE

Plan view of top face

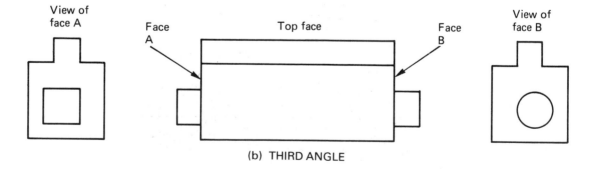

View of
face A

Face
A

Top face

Face
B

View of
face B

(b) THIRD ANGLE

Figure 9.13 Orthographic projection

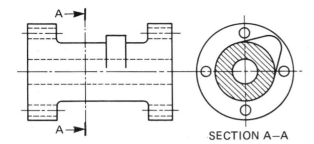
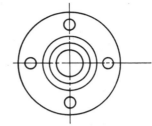
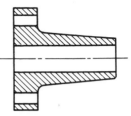

SECTION A—A

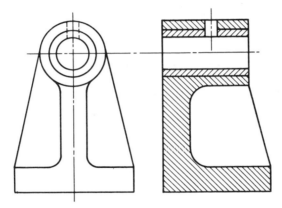

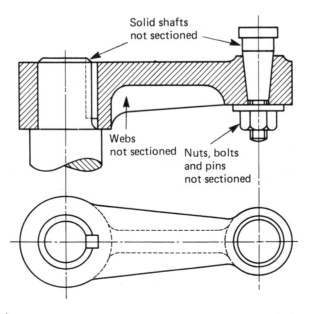

Solid shafts
not sectioned

Webs
not sectioned

Nuts, bolts
and pins
not sectioned

Figure 9.14 Section drawings

Types of line		
Line	Description	Application
A  ———————	Continuous thick	A1 Visible outlines A2 Visible edges
B  ———————	Continuous thin	B1 Imaginary lines of intersection B2 Dimension lines B3 Projection lines B4 Leader lines B5 Hatching B6 Outlines of revolved sections B7 Short centre lines
C  D 	Continuous thin irregular Continuous thin straight with zigzags	*C1 Limits of partial or interrupted views and sections, if the limit is not an axis †D1 Limits of partial or interrupted views and sections, if the limit is not an axis
E ■ ■ ■ ■ ■ F  – – – – –	Dashed thick Dashed thin‡	E1 Hidden outlines E2 Hidden edges F1 Hidden outlines F2 Hidden edges
G  — – — – —	Chain thin	G1 Centre lines G2 Lines of symmetry G3 Trajectories and loci G4 Pitch lines and pitch circles
H 	Chain thin, thick at ends and changes of direction	H1 Cutting planes
J  ■■ – ■■ – ■■	Chain thick	J1 Indication of lines or surfaces to which a special requirement applies (drawn adjacent to surface)
K  — – – — – – —	Chain thin double dashed	K1 Outlines and edges of adjacent parts K2 Outlines and edges of alternative and extreme positions of movable parts K3 Centroidal lines K4 Initial outlines prior to forming §K5 Parts situated in front of a cutting plane K6 Bend lines on developed blanks or patterns

NOTE. The lengths of the long dashes shown for lines G, H, J and K are not necessarily typical due to the confines of the space available.

†This type of line is suited for production of drawings by machines.

‡ The thin F type line is more common in the UK, but on any one drawing or set of drawings only one type of dashed line should be used.

§ Included in ISO 128-1982 and used mainly in the building industry.

Figure 9.15 Types of line
(Extracts from BS 308 are reproduced by permission of BSI. Complete copies of the standard can be obtained from them at Linford Wood, Milton Keynes, Bucks MK14 6LE.

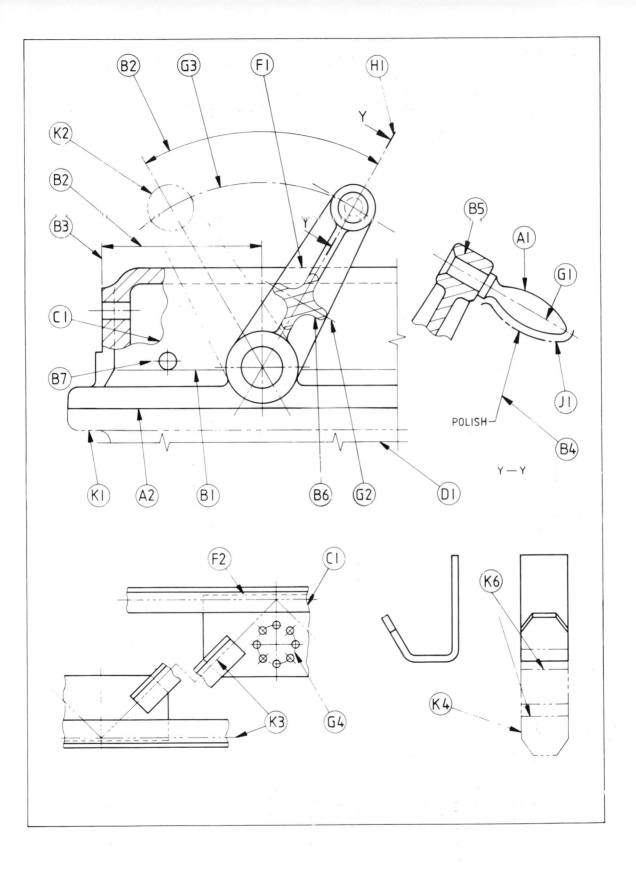

POLISH

Y — Y

TITLE	SUBJECT	CONVENTION

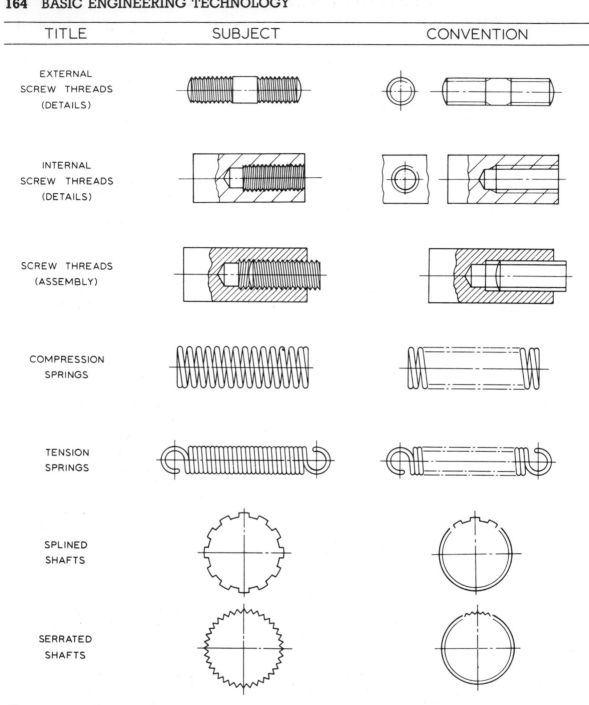

EXTERNAL SCREW THREADS (DETAILS)

INTERNAL SCREW THREADS (DETAILS)

SCREW THREADS (ASSEMBLY)

COMPRESSION SPRINGS

TENSION SPRINGS

SPLINED SHAFTS

SERRATED SHAFTS

Figure 9.16 Conventions for representation of common features

TITLE	SUBJECT	CONVENTION

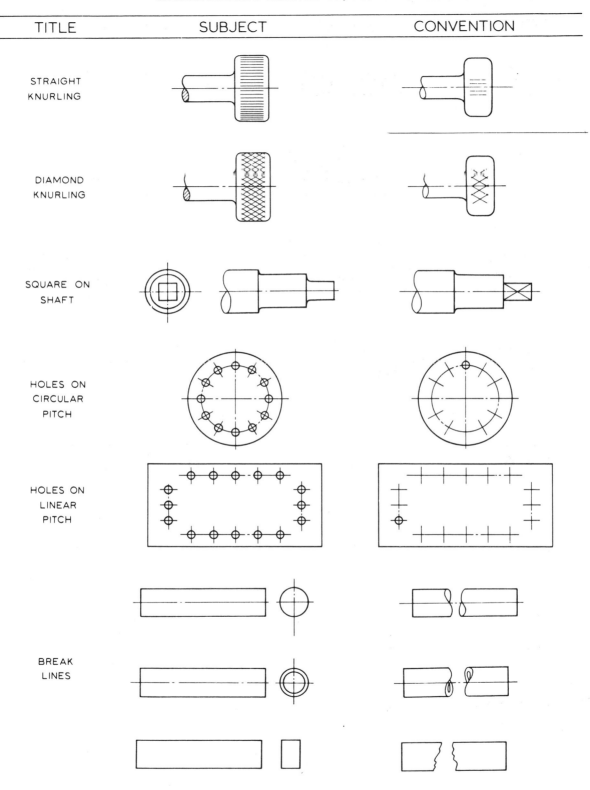

STRAIGHT
KNURLING

DIAMOND
KNURLING

SQUARE ON
SHAFT

HOLES ON
CIRCULAR
PITCH

HOLES ON
LINEAR
PITCH

BREAK
LINES

9.7 Dimensioning

When a component is being dimensioned, the dimension lines and the projection lines should be thin full lines and, where possible, they should be placed outside the outline of the object as shown in Figure 9.17a. They should be half the thickness of the outline. There should be a small gap between a projection line and the outline, and the projection line should extend to just beyond the dimension line. The purpose of these rules is to allow the outline of the object to stand out prominently from the other lines and to prevent confusion. Dimension lines end in an arrow which should touch the projection line to which it refers. All dimensions should be placed in such a way that they can be read from the bottom right-hand corner of the drawing.

Figure 9.17b shows the effect of chain dimensioning a series of holes or other features. Although the designer intends a tolerance of ± 0.2 mm, the cumulative error of the final (right-hand) hole position can be as great as ± 0.6 mm. In this situation it is better to dimension from a common datum as shown in Figure 9.17c. This prevents the build-up of cumulative errors.

Further examples of dimensioning techniques are shown in Figure 9.18.

It has already been stated in Chapter 4 that it is impossible to work or measure exactly to a given size, and that the designer must provide limits of size between which the component will be acceptable. These may be general limits, or the limits may be applied specifically to a given dimension.

9.8 Pictorial views

Perspective drawings are rarely used by engineers; pictorial views are usually drawn in oblique or isometric projection.

Oblique view

An example is shown in Figure 9.19a. The elevation is drawn true shape and true size as in orthographic drawing. For simplicity, the face chosen for the elevation should contain any arcs or circles, as these can be drawn with compasses.

The lines disappearing back into the drawing (the receders) are drawn at 45° and only half length.

Isometric view

An example is shown in Figure 9.19b. All receders are at 30° to the horizontal, and all lengths are usually drawn true size. The receders are sometimes drawn to isometric scale, but the advantages in appearance do not justify the time and cost involved. Although the isometric view is more pleasing in appearance than the oblique view, all curves in an isometric drawing have to be constructed as ellipses, as shown in Figure 9.19c.

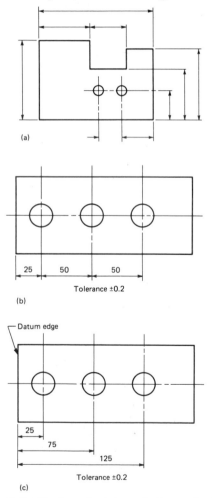

Figure 9.17 Basic principles of dimensioning: (a) projection and dimension lines (b) chain dimensioning (c) dimensioning from a datum

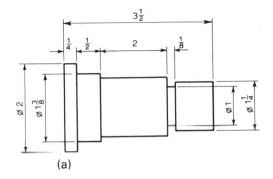

(a)

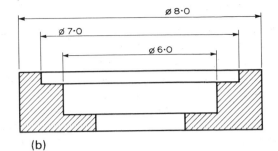

(b)

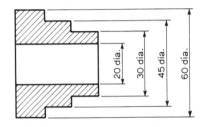

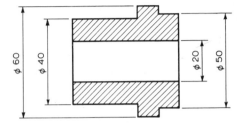

(c)

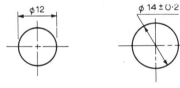

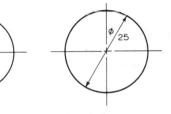

(d)

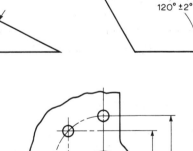

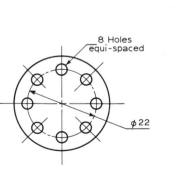

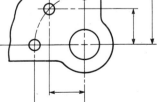

(e)

Figure 9.18 Further examples of dimensioning: (a) overall dimensions outside intermediate dimensions (b) dimensions staggered to avoid confusion (c) alternative methods of indicating diameters (d) dimensioning circles and angles (e) positioning of holes

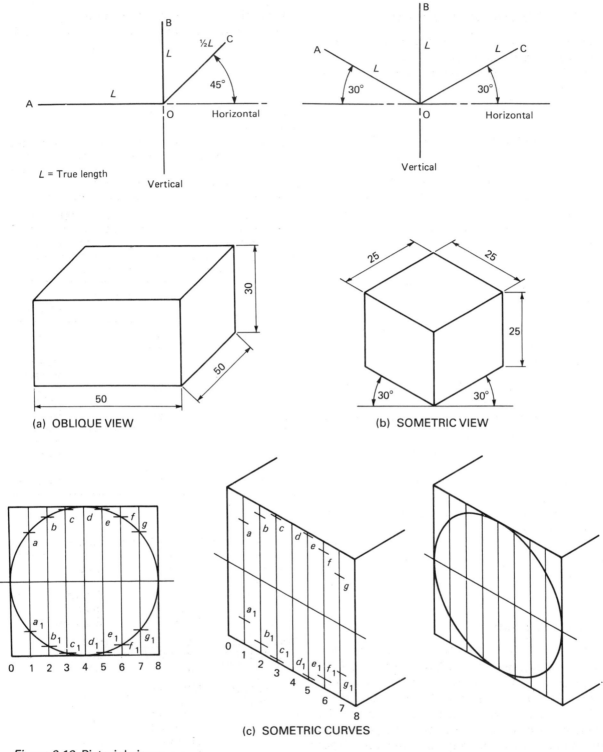

L = True length

(a) OBLIQUE VIEW

(b) SOMETRIC VIEW

(c) SOMETRIC CURVES

Figure 9.19 Pictorial views

9.9 Identification of components

Figure 9.20 shows an exploded pictorial view. This type of drawing is frequently used in service manuals for identifying components and showing the sequence for assembly.

9.10 Interpreting tables and graphs

Frequently a drawing or a specification may call up materials, components or data with which the engineer is unfamiliar. For example, it may call up an M12 hexagon steel bolt to secure a particular component. This information is incomplete, and the engineer will need to refer to manufacturers' data sheets or to workshop pocket books to find:

(a) The pitch of the thread (fine or coarse series)
(b) The tapping drill size
(c) The preferred lengths, so that the bolt will have a full length of thread in engagement with the mating component, yet not have surplus length protruding from the hole.

Again, a drawing may call up sheet metal thickness as 14 SWG (standard wire gauge). Reference to tables will show that the sheet will have a thickness of 2.03 millimetres (0.080 inches).

It has already been stated that only graphs which obey a mathematical expression can be interpolated. Figure 9.21a shows the relationship between current and potential for a conductor of given resistance, together with some interpolated values. Figure 9.21b shows the relationship between twist drill diameter and spindle speed for a given cutting speed, together with some interpolated values.

9.11 Colour coding

Colour coding is used for a number of purposes, and some examples are included in this section. Your safety and the safety of others depend on a clear understanding and use of these codes.

Figure 9.22 shows the method of identifying the contents of pipes and conduits. Table 9.1 shows the method of identifying the contents of gas cylinders. Table 9.2 shows the method of identify-ing electric cables. Figure 9.23 shows how colour coding is used to indicate the ohmic value and tolerance of carbon compound resistors.

Table 9.1 Colour coding for gas cylinders (BS 349: 1932)

A cylinder which is coloured wholly red or maroon, or has a red band around it near the top, contains a combustible gas. A cylinder having a yellow band around its top contains a poisonous gas. The table indicates the colour coding used for various gases.

Whilst on site the cylinders should be kept clean so that they can be easily identified. On no account must the colour of any cylinder be changed by any person either on site or in the workshop. Acetylene cylinders must never be stacked horizontally.

Gas	Ground colour of cylinder	Colour of bands
Acetylene	Maroon	None
Air	Grey	None
Ammonia	Black	Red and yellow
Argon	Blue	None
Carbon monoxide	Red	Yellow
Coal gas	Red	None
Helium	Medium brown	None
Hydrogen	Red	None
Methane	Red	None
Nitrogen	Dark grey	Black
Oxygen	Black	None

Table 9.2 Identification of electrical cables

Service	Cable	Colour
Single phase Flexible	Live	Brown
	Neutral	Blue
	Earth	Green/yellow
Single phase Non-flexible	Live	Red
	Neutral	Black
	Earth	Green
Three phase Non-flexible	Line (live) (colour denotes phase)	Red White Blue
	Neutral	Black
	Earth	Green

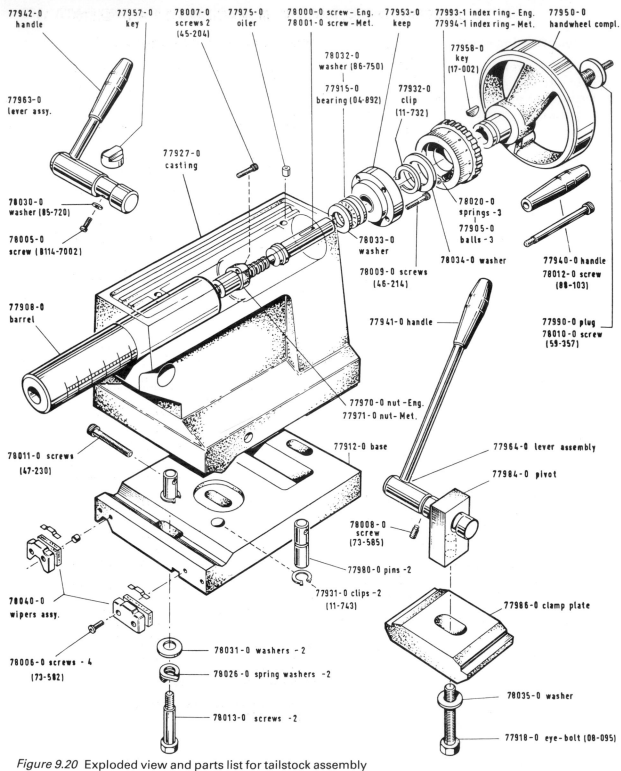

Figure 9.20 Exploded view and parts list for tailstock assembly
(Reproduced by permission of the Colchester Lathe Co. Ltd.)

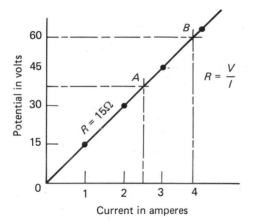

(a) Voltage/current graph:
   A: current when $V = 37.5$ volts is 2.5 amperes
   B: potential when $I = 3.8$ amperes is 57.0 volts

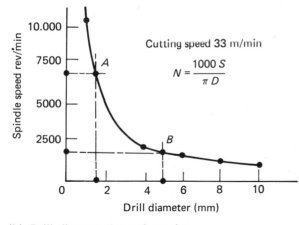

(b) Drill-diameter/speed graph:
   A: spindle speed for 1.5 mm drill is 7000 rev/min
   B: spindle speed for 5.0 mm drill is 2100 rev/min

Figure 9.21 Interpretation (interpolation) of graphs

COLOUR CODE IN ONE OF FIVE WAYS

(a) WHOLE PIPE PAINTED

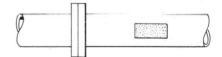

(b) PANEL PAINTED

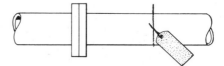

(c) COLOURED LABEL ATTACHED

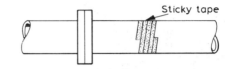

(d) COLOURED STICKY TAPE FASTENED

(e) COLOURED BANDS PAINTED

Colour	Contents
White	Compressed air
Black	Drainage
Dark grey	Refrigeration and chemicals
Signal red	Fire
Crimson or aluminium	Steam and central heating
French blue	Water
Georgian green	Sea, river and untreated water
Brilliant green	Cold water services from storage tanks
Light orange	Electricity
Eau-de-Nil	Domestic hot water
Light brown	Oil
Canary yellow	Gas

Figure 9.22 Colour codes for the contents of pipes (from BS 1710)

1st number
2nd number
number of zeros
tolerance

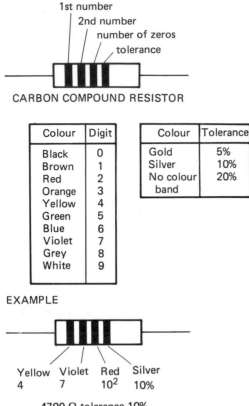

CARBON COMPOUND RESISTOR

Colour	Digit
Black	0
Brown	1
Red	2
Orange	3
Yellow	4
Green	5
Blue	6
Violet	7
Grey	8
White	9

Colour	Tolerance
Gold	5%
Silver	10%
No colour band	20%

EXAMPLE

Yellow	Violet	Red	Silver
4	7	10^2	10%

4700 Ω tolerance 10%

Figure 9.23 Resistor colour code

Exercises

For each exercise, select *one* of the four alternatives.

1 Three-dimensional solid objects are usually represented on a detail drawing using
   (a) oblique projection
   (b) orthographic projection
   (c) isometric projection
   (d) axonometric projection.
2 Drawings for service engineers must show the relationship between the assembled parts clearly and in such a way that the parts can be quickly identified by using
   (a) first-angle orthographic projection
   (b) third-angle orthographic projection
   (c) perspective drawing
   (d) exploded views.

3 Figure 9.24 has been drawn in
   (a) first-angle orthographic projection
   (b) third-angle orthographic projection
   (c) oblique projection
   (d) isometric projection.

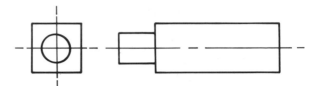

Figure 9.24

4 Figure 9.25 has been drawn.
   (a) first-angle orthographic projection
   (b) third-angle orthographic projection
   (c) oblique projection
   (d) isometric projection.

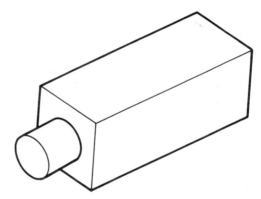

Figure 9.25

5 Figure 9.26 shows one method of presenting statistical data. It is called a
   (a) bar chart
   (b) histogram

1985

1986

1987

= 100 pupils

Figure 9.26

(c) pie chart
(d) pictogram (ideograph).
6 Figure 9.27 shows a
  (a) bar chart
  (b) histogram
  (c) pie chart
  (d) pictogram (ideograph).

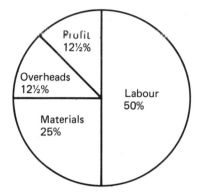

Figure 9.27

7 Figure 9.28 shows the relationship between current and potential for a particular conductor. When a current of 2.5 amperes is flowing, the corresponding potential will be
  (a) 5 volts
  (b) 15 volts
  (c) 25 volts
  (d) 35 volts.
8 Given that the resistance in ohms of a conductor equals the potential in volts divided by the current flowing in amperes, the resistance of

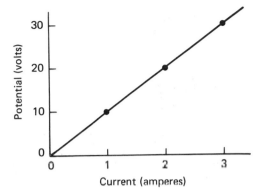

Figure 9.28

the conductor used in obtaining the data for Figure 9.28 is
  (a) 0.1 ohms
  (b) 1.0 ohms
  (c) 10.0 ohms
  (d) 100.0 ohms.
9 Figure 9.29 represents a simple radio receiver. This type of diagram is called
  (a) an assembly drawing
  (b) a block diagram
  (c) a flow chart
  (d) a location drawing.
10 BS 308: Part 1 shows a number of representations of commonly used drawing details such as screw threads in highly simplified form. These are referred to as
  (a) conventions
  (b) abbreviations
  (c) symbols
  (d) standards.

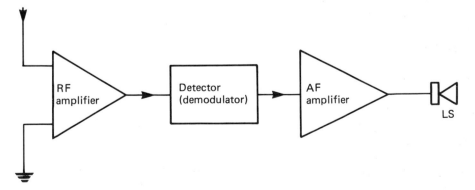

Figure 9.29

11 In Figure 9.30 the line A is referred to as
   (a) a dimension line
   (b) a leader line
   (c) an outline
   (d) a projection line.

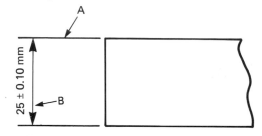

Figure 9.30

12 In Figure 9.30 the line B is referred to as
   (a) a dimension line
   (b) a leader line
   (c) an outline
   (d) a projection line.
13 The shaded portion of Figure 9.31 represents
   (a) a section
   (b) a revolved section
   (c) another girder at right angles to the first
   (d) an end view.

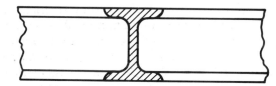

Figure 9.31

14 If a pipe is coloured yellow it contains
   (a) cold water

   (b) hot water
   (c) steam
   (d) natural gas.
15 If a gas cylinder is coloured maroon it contains
   (a) propane
   (b) butane
   (c) oxygen
   (d) acetylene.
16 The insulation colours of a three-core flexible cable for live, neutral and earth are respectively
   (a) blue; brown; green/yellow striped
   (b) brown; blue; green/yellow striped
   (c) brown; green/yellow striped; blue
   (d) blue; green/yellow striped; brown.

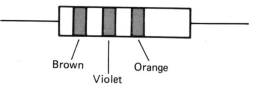

Brown   Violet   Orange

Figure 9.32

17 The resistor shown in Figure 9.32 has a nominal resistance value of
   (a) 370 ohms
   (b) 17,000 ohms
   (c) 173 ohms
   (d) 371 ohms.
18 If a silver band was added to the resistor shown in Figure 9.32, the tolerance would be
   (a) 5 per cent
   (b) 10 per cent
   (c) 15 per cent
   (d) 20 per cent.

10

Assembly and dismantling

10.1 The need for assembly

The purpose of assembly is to put together a number of individual components to form a whole device, structure or system. To achieve this aim, attention must be paid to the following key factors:

Sequence of assembly This must be planned so that as each component is added, its position in the assembly and the position of its fastenings are accessible.

Technique of joining This must be selected to suit the components being joined, the materials from which they are made, and their service function. Examples of different joining techniques were considered in Chapter 8.

Position of joints In designing a new assembly, or planning a modification to an existing assembly or installation, the position of any joints must be preplanned for the convenience not only of assembly but also of maintenance and re-placement.

Interrelationship and identification of parts Identification of parts and their position in an assembly can usually be determined from assembly drawings or exploded view drawings, as discussed in Chapter 9. Interrelationship markings are often included on components. For example, the various components and joints of structural steelwork are given number and letter codes to facilitate identification and assembly on site. Again, when a new timing belt is fitted to a car engine the engine block is marked and so are the crankshaft and camshaft pulleys. These marks must be in alignment when the belt is fitted to ensure that the camshaft is properly timed.

Tolerances The assembly technique must take into consideration the accuracy and finish of the components being assembled. There is a considerable difference in the quality of workmanship, the techniques selected and the tools used in the assembly of structural steelwork, for example, compared with the assembly of a precision tool or an artificial satellite.

Protection of parts Components awaiting assembly require protection against damage and corrosion. In the case of structural steelwork this may merely consist of painting with a red oxide primer and careful stacking. Precision components will require treating with an anti-corrosion compound such as lanolin which can be easily removed at the time of assembly. Bores should be sealed with plastic plugs and screw threads with plastic caps. Precision-ground joint surfaces and datum surfaces must be protected against damage. Heavy components must be provided with suitable eye-bolts so that precision surfaces are not damaged by the use of slings during lifting. Complex and vulnerable assemblies, such as aircraft engines, should be mounted in suitable cradles to prevent damage and distortion during transit and whilst awaiting assembly into the aircraft.

10.2 The need for dismantling

Dismantling must be considered with the same care as assembly if the individual components are not to be damaged or destroyed.

Periodic inspection and maintenance Many assemblies have to be opened up for inspection and maintenance. For example: the reduction gears between the turbines and propellers of ships, steam boilers require internal inspection for corrosion; and internal combustion engines require the cylinder heads to be removed from time to time for inspection of the cylinder bores, decarbonizing and valve grinding. In all these examples care must be taken in opening up the assemblies if damage to the joint faces is to be avoided so that oil- and gas-tight joints can be made on reassembly.

Repair It is often more convenient to repair components in the workshop than on site, and in any case the component will usually have to be removed for improved accessibility. Careful dismantling is required to remove the failed component.

Replacement of worn components Provision has to be made in assemblies for the easy dismantling and removal of components subject to wear. For example, it must be possible to remove the circulating pump from a central heating system without draining down the system and dismantling the pipework. To this end, full-flow stop valves are fitted each side of the pump and flanged or union joints are provided between the pump and the pipework.

Trial assembly Some assemblies are too large to transport to site completely built up. On the other hand it would be extremely embarrassing and costly if large subassemblies did not fit together after delivery to site. Therefore a trial assembly in the erection shop will allow corrections and adjustment to be made so that, after dismantling and delivery, easy site assembly is ensured.

10.3 Methods of assembly and dismantling

On-site assembly: no prefabrication

Except for small alterations, it is rare to carry out assembly on site without some prefabrication. However, an example of work without prefabrication is shown in Figure 10.1, where a pipe fitter has to fabricate and fit a bracket to support a long pipe run. The bracket would be bent up and

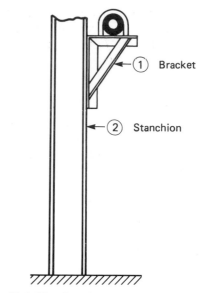

Figure 10.1 Pipe bracket

welded out of angle iron by the fitter and drilled and bolted to the stanchion. There would be no drawings prepared and the work would be carried out on the initiative of the pipe fitter.

On-site assembly: partial fabrication

Fabricated and manufactured components are often used to build up systems which are assembled and fitted on site. An example is the electrical installation in a building. The switchgear, fuseboards, conduit fittings, accessories etc. are delivered on site completely finished and ready for assembly. The conduit, however, is delivered in standard straight lengths and has to be cut, threaded, bent and fitted on site and coupled to the switchgear etc. ready for wiring up and completion of the system.

Large-structure assembly

This includes the assembly of structures such as ships, oil rigs, bridges, cranes etc. Many of the components are extremely heavy and large cranes are used. In addition, there are many teams of fitters at work on different parts of the site at the same time. Therefore the work has to be carefully planned so that it progresses in the correct sequence and without the teams getting in each other's way. The work is potentially hazardous because of its size and weight and great care has to be taken by individual workers, not only for their own safety but for the safety of other groups on the site.

Trial assembly

Many large assemblies such as turbine-generator sets for power stations are too large for transport from the maker's factory to the site completely built up. However, because they need to be assembled and tested before delivery on site, they are given a trial erection in the factory. When assembly has been proved satisfactory, the structures are partially dismantled into a series of subassemblies which can be easily reassembled on site after transportation.

One-off assembly

An example of this type of assembly would be the building up of a press tool by the fitter from the components supplied from the machine shop and by reference to the assembly drawing. Some of the components will be supplied only partly finished and the fitter will have to position them correctly, drill and tap screw holes, and drill and ream dowel holes for location as assembly proceeds. In this case the fitter has a considerable influence on the accuracy and performance of the finished tool.

Batch assembly

This refers to the assembly of a small number of devices of the same type. An example would be the building up of a batch of coolant pumps. The components and subassemblies (motor, body, impeller, fixings etc.) would be supplied fully machined and ready for assembly. The fitter would only be expected to assemble the components together in the correct sequence without any adjustment or additional finishing.

Line assembly

This is used in the mass production of products such as motor cars. No single highly skilled, fitter completely builds a car. Instead a large number of assembly line workers of limited skill operate at stations along an assembly line. Each contributes a particular component or subassembly to the vehicles as they slowly travel past on the conveyor belt. The components have to be completely finished to a high degree of accuracy, since there is no time for adjustment or correction. Because of the repetitive nature of the work, robots are taking over many of the assembly functions on the assembly line.

On-site dismantling and repair

An example of this is a leak in a pipe joint. The system is drained and the joint is dismantled. The cause of the failure is identified (usually a failed gasket between the flanges) and a replacement component is fitted. The system is then refilled and tested under pressure.

On-site dismantling: workshop repair

More complex repairs need to be carried out in the workshop where special facilities are available. An example is the rewinding of the coils of an electric motor which has burnt out. The motor is dismantled from the equipment it is driving, returned to the workshops of a firm specializing in rewinds, repaired and returned to site, where it is reinstalled.

On-site replacement and works reconditioning

In the previous example the equipment driven by the motor would be left standing whilst the motor was repaired. This can be overcome by fitting a reconditioned replacement motor as soon as the faulty motor is removed, and then repairing and keeping the original motor as a spare for use elsewhere when another breakdown occurs.

10.4 Selection of methods of assembly and dismantling

The methods of assembly and dismantling chosen will be influenced by a number of factors, some of which will now be considered.

Size

The size of the system or the size of the individual components in an assembly will affect both assembly and dismantling. A small bush may be assembled into its housing or driven out of its housing using a hammer and drift. A larger bush would require to be assembled or dismantled using a hydraulic press. Whilst small items of equipment such as portable hand tools can be stripped and rebuilt on the bench, large machines will require cranes and special facilities. On no account should work be carried out under equipment suspended by a crane; the equipment should be lowered on to suitable supports.

Static and moving parts

The assembly of a static structure such as structural steelwork for a building completes the operation, and providing the component girders have been properly prefabricated and the fixing holes properly aligned no faults should occur. However, in a complex assembly involving moving parts, such as the gearbox of a machine tool, not only must the individual components be correctly assembled with great care to avoid damage which could cause misalignment, but at each stage of assembly the components must be checked for correct movement and smooth running.

Accuracy

Where the level of accuracy is low, the skills involved in assembly are much less than where the level of accuracy is high. A jobbing gardener can successfully fit a new wheel to his wheelbarrow, but it takes a highly skilled fitter to correctly fit and scrape in the bearing shells of a large plain journal bearing such as those used to support the propeller shaft of a large ship. Greater care and the use of

special equipment is required when dismantling assemblies made to high levels of accuracy than when only coarse tolerances are involved. For example, a sacktruck wheel can be removed by simply removing the split pin and withdrawing the washer and wheel by hand. In contrast, an accurately fitted pulley wheel requires the use of a special extractor, as shown in Figure 10.2, to draw it off the shaft without damage to the mating surfaces.

Accessibility

Wherever possible, fixings should be accessible for easy assembly and dismantling using standard tools. However, the constraints of design are such that sometimes special tools have to be devised. Further, where a complex assembly is built up stage by stage it is essential to carry out the work according to a planned sequence so that all the fixings at each stage are accessible. Similarly, when dismantling the assembly, it might be necessary to remove portions of the assembly which are working normally to reach components which have failed.

Number of parts

Where a large number of parts are required in an assembly, they should be checked off against a parts list before work commences to ensure that they are all available. They should be laid out on a clean and clear bench in the sequence in which they will be required, and when assembly is complete all the components should have been used, with none surplus and none short, as a check of correct assembly.

Similarly the greater the number of parts the greater will be the complexity of the dismantling operation. It is highly unlikely that all the parts will fail at the same time, and usually only one of the subassemblies which make up the whole system will need to be dismantled. For example, in a machine tool the hydraulic pump may have to be stripped down and repaired. Since the pump consists of a number of parts it is essential to label or identify the parts and mark them so that they can be reassembled the correct way round, as shown in Figure 10.3.

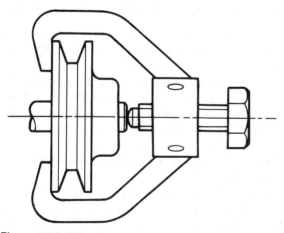

Figure 10.2 Wheel extractor

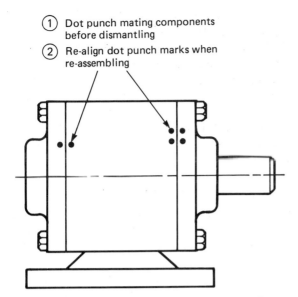

① Dot punch mating components before dismantling

② Re-align dot punch marks when re-assembling

Figure 10.3 Identification and alignment of components using dot punch marks before dismantling and on reassembling

Transport

Where equipment has to be dismantled for transporting to a different site, minimum disturbance and dismantling should be aimed for. However, the loading facilities and the capacity of the transport must be taken into account. It may be convenient to crane the dismantled assemblies on to a large low-loader lorry at the factory, but this is of no use if access to the site is restricted and the crane facilities limited.

Nature of repair

The method of dismantling will be affected by the nature of the repair. Greater care must be taken if the failed components are to be reclaimed and reused than if they are to be scrapped and replaced. Frequently fixings become immovable by ordinary dismantling techniques when they have been undisturbed for many years.

On-site assembly conditions

As for transport, the extent of dismantling is often controlled by the site conditions where the plant is to be reassembled. For example, there may be restrictions on access, headroom and crane ca-

pacity, on the load-bearing capability of the ground, and so on.

Ease of dismantling

It has already been stated that when joints have been undisturbed for a long time they may become corroded and difficult to dismantle by conventional methods. Screwed fastenings may require the application of a penetrating oil; in extreme cases, heating up the nut with a gas torch will cause it to expand and the seal will be broken. If the head of the bolt has to be chiselled off, care must be taken that the force used does not damage the rest of the assembly. Where riveted joints are involved, these may have to be carefully centred and drilled out or, again, the heads may have to be chiselled off.

10.5 Relationship between assembled components

There are a number of different relationships between assembled components depending on their function. Some examples of the different relationships which can exist between mating components are shown in Figure 10.4.

10.6 Forces in assembly and dismantling

It is important when assembling and dismantling screwed fastenings that they are not overstressed. When the nut is tightened on a bolt, the bolt stretches slightly and acts as a powerful spring as it tries to shrink back to its original length, and thus holds the joint tightly together. If the bolt is overtightened it takes a permanent set and loses its elasticity. This not only weakens the bolt, but destroys the spring action which keeps the joint tightly closed.

When using a hammer or mallet it is important to make sure that the component is strong enough to withstand the blow. The weight of the hammer should be selected accordingly.

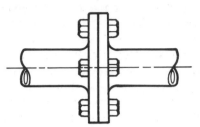

(a) FLANGED COUPLING: ASSEMBLED PARTS DO NOT MOVE RELATIVE TO EACH OTHER AND THE JOINT IS FLUID TIGHT

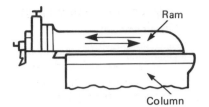

(b) SHAPING MACHINE RAM AND COLUMN: ONE COMPONENT (RAM) SLIDES RELATIVE TO THE OTHER (COLUMN)

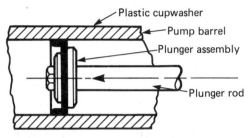

(c) LOW-PRESSURE PUMP: ONE COMPONENT (PLUNGER) SLIDES RELATIVE TO THE OTHER (CYLINDER) BUT IS FLUID TIGHT

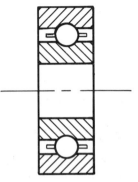

(d) ANTI-FRICTION (BALL) BEARING: ONE COMPONENT (INNER RACE) ROTATES RELATIVE TO THE OTHER (OUTER RACE)

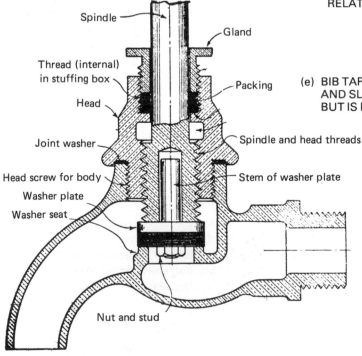

(e) BIB TAP: ONE COMPONENT (SPINDLE) ROTATES AND SLIDES RELATIVE TO THE OTHER (HEAD) BUT IS FLUID TIGHT

Figure 10.4 Relationship between assembled components

10.7 Tools used for assembly and dismantling

Figure 10.5 shows various types of spanner and key used in assembly and dismantling operations. Spanners are proportioned so that the length of the spanner provides sufficient leverage for a person of average strength to tighten the fastening correctly. On no account should the spanner be extended with a tube or other device, as this will not only overstress the bolt but also strain the jaws of the spanner so that it will not fit the nut properly. This can lead to damage to the nut and injury to the fitter if the spanner slips.

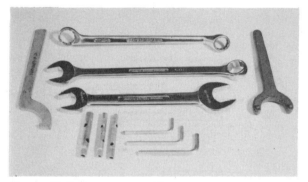

Figure 10.5 Spanners and keys

Adjustable spanners or 'monkey wrenches' should only be used as a last resort. The lack of rigidity in the jaws leads to damage to the nut. In addition, the fact that the adjustable spanner is a fixed length means that small nuts and bolts are easily overtightened and sheared off, and large nuts and bolts are inadequately tightened.

Where the force used to tighten a screwed fastening is critical, a torque spanner is used (Figure 10.6). The force to be exerted on the fastening is specified by the designer and the fitter sets the torque spanner to this value. When the fastening is tightened correctly a clutch in the head of the spanner slips and no further force can be exerted on the bolt or nut.

Screwdrivers must also be chosen to fit the head of the screw, otherwise the slot will be damaged. Pliers should not be used to tighten small nuts; the corners of the nut will be damaged, and it will then not be possible to use a spanner on it.

Hammers are also used for assembly and dismantling. Care should be taken that the components being struck are not bruised and damaged. Soft-faced hammers should be used where possible, or a soft drift should be placed between the hammer and the component being struck. A sharp blow will often loosen a joint when steady pressure has had no effect.

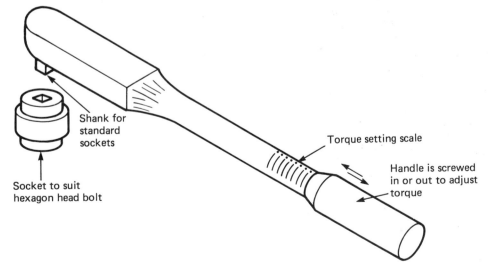

Figure 10.6 Torque spanner

10.8 Miscellaneous equipment used in assembly and dismantling

Assembly

Figure 10.7 shows a number of different types of seal and some typical applications. Seals are normally used to keep mechanical joints fluid tight and to prevent dirt getting into the joint. They are designed so that the pressure tends to keep the seal closed on the joint. For engineering purposes, seals are normally made from a synthetic rubber such as neoprene which is wear resistant and is unaffected by oil.

Thermal compression jointing requires equipment to expand the outer component or to shrink the inner component. One of the most common ways of expanding large components is by a ring of gas jets arranged around the component. Small components can be heated up in a muffle furnace. Induction heating can also be used. The use of coolants such as solid carbon dioxide or liquid nitrogen to supercool and shrink the inner component was discussed in Section 8.4.

Figure 10.8 shows an example of a manually operated press and a simple hydraulic press. These are used when fitting ball and roller bearings and pressing in bushes.

Measuring and checking equipment used in assembly and dismantling is the same as that described in Chapter 4 for manufacturing purposes.

Dismantling

Generally, the same tools are used for dismantling as are used for assembly. However, there are some additional requirements:

(a) Cleansing agents such as pressure steam jets, paraffin or solvents (for example, trichloroethylene) are used to clean components after dismantling and before repair work is commenced. Since solvents are toxic and flammable, great care must be exercised in their use.

(b) Dyes and markers are used to identify components as they are dismantled so that they

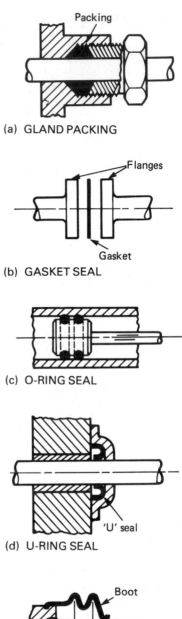

(a) GLAND PACKING

(b) GASKET SEAL

(c) O-RING SEAL

(d) U-RING SEAL

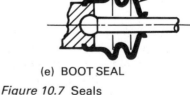

(e) BOOT SEAL

Figure 10.7 Seals

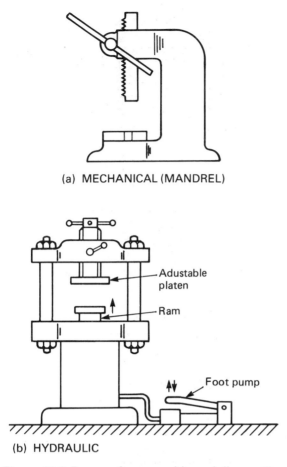

(a) MECHANICAL (MANDREL)

(b) HYDRAULIC

Figure 10.8 Presses for assembly and dismantling

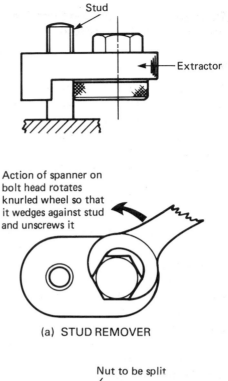

(a) STUD REMOVER

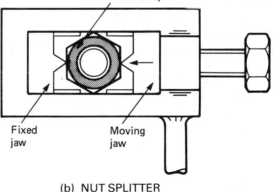

(b) NUT SPLITTER

Figure 10.9 Removal of studs and nuts

can be reassembled in the correct sequence. Paint sprays, stencils, centre punches and number punches are all used for this purpose. An example was shown in Figure 10.3.

(c) Penetrating oils are used to assist in loosening rusted components. They are not suitable for lubrication.

(d) Studs can be extracted by locking two nuts together on the exposed end and applying the spanner to the lower nut so that it tightens against the upper nut as the stud is removed.

Alternatively, a stud remover can be used. This has an eccentrically mounted hardened knurled wheel which is wedged against the stud by the action of the spanner, as shown in Figure 10.9a. The greater the pull on the spanner, the tighter the knurled wheel grips the stud.

Nut splitters consist of a pair of hardened jaws which cut into the nut as the bolt is tightened, as shown in Figure 10.9b.

Broken screws and studs can be removed using an 'easy-out'. This is a tapered tap with a left-handed coarse-pitch thread. A hole is drilled into the broken screw and the easy-out is screwed in. Since the thread of this tool is left-handed, the action of screwing it in turns the broken screw in the correct direction to remove it.

(e) Special-purpose spanners and keys are required where production and design considerations have prevented standard fixings being used. These are normally provided with the machine in the first instance for maintenance purposes, and their availability should be confirmed before dismantling commences.

(f) Gear pullers and pulley wheel pullers are required for the removal of these components from shafts. An example was shown in Figure 10.2.

(g) Flame cutting is widely used in the dismantling and demolition of structural steelwork and pipework for site reclamation ready for the installation and assembly of new plant.

10.9 Precautions

Assembly or dismantling should not be undertaken without first considering the hazards associated with a particular job. Some of these will now be discussed.

Forces

The hazards associated with moving heavy loads have already been considered in Chapter 3. The forces acting on a piece of equipment must be assessed before dismantling commences. Unfastening the flanged joints in Figure 10.10a will leave the steam valve unsupported so that it will fall. Lifting equipment must be provided to take the weight of the valve before dismantling commences. Care must also be taken that the forces used to tighten or loosen fixings do not dislodge the assembly being worked upon. Containers should be drained before moving, as surging liquids within the containers can make them unstable (Figure 10.10b).

Contents

Before dismantling any pipework system, the contents of the system must be ascertained. The

substances in the pipes may be flammable, toxic or under high pressure. The BS colour code (Figure 9.22) helps in identification of the contents of the system. The system will most likely have to be purged and a 'permit-to-work' obtained before work can commence. Closing down the system and rendering it safe is a senior supervisory responsibility and should not be attempted by the fitter. Valves should be locked so that they cannot be opened whilst work is in progress.

Electricity

Assembling and dismantling machines and plant driven by electric motors or involving the use of heavy-duty electrical heating systems should not be attempted until the equipment has been isolated by a qualified electrician. The fuses should be drawn and the switches locked off. The maintenance engineer should keep the key and the fuses until the work is complete.

Temperature

Pumps, pipework and valves operating at high temperatures may have to receive emergency repairs before there is time for them to cool down. Suitable protective clothing must be worn, and respirators may be required. The maintenance fitter should be closely supervised so that he can be rescued if overcome by the heat.

Chemicals

Codes of practice must be rigorously observed when working on plant which is used in conjunction with dangerous and radioactive chemicals. The work must only be carried out under the supervision of authorized and qualified persons. Appropriate protective clothing must be worn. Precautions must be taken against accidental leakage of such dangerous substances, which could cause a health hazard over a large area.

10.10 The influence of material properties

Components and fixings used in assembly and dismantling operations are generally specified by the design engineer. However, care must be taken to recognize the properties of these materials so

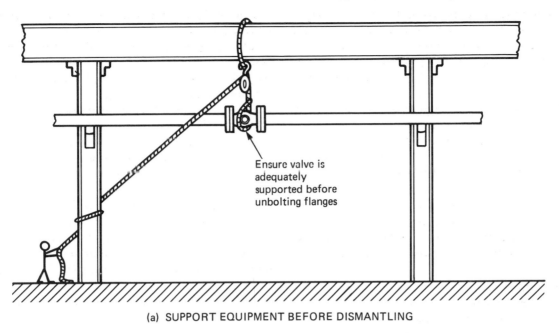

Ensure valve is
adequately
supported before
unbolting flanges

(a) SUPPORT EQUIPMENT BEFORE DISMANTLING

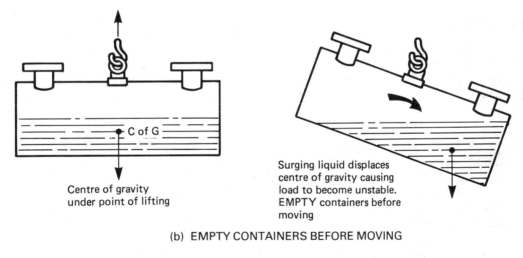

C of G

Centre of gravity
under point of lifting

Surging liquid displaces
centre of gravity causing
load to become unstable.
EMPTY containers before
moving

(b) EMPTY CONTAINERS BEFORE MOVING

Figure 10.10 Precautions when dismantling and moving loads

that the components are not damaged by unsuitable treatment. For example, cast iron components are brittle and should not be driven into position with a hard-faced hammer; a hide-faced mallet would be more appropriate. Although a design consideration, the material for a particular jointing technique must have suitable mechanical properties. For example, cast iron is suitable for the inner component of a shrink or compression joint since it is strong in compression; it would fail if used for the outer component since it is weak in tension. Steel or some similar high-tensile material would be required for the outer component of this type of joint.

When re-erecting an assembly after repair, it is essential that any new fixings which are used are to the same specification as the originals. If a particular size and type of high-tensile bolt and nut was used originally, then an identical type of bolt and nut must be used as a replacement. A serious accident could occur if a bolt and nut of weaker specification was used for convenience.

Sealing agents must be appropriate for the joint being made. They must be capable of withstanding the temperatures and pressures involved and should not be dissolved by the fluids flowing in the pipes.

Corrosion should be minimized by careful choice of metals for fixings. These should be the same as the metal of the components being joined: for example, steel rivets or bolts should be used for connecting steel components, and copper rivets or screws for copper components. This prevents electrolytic corrosion, which occurs when dissimilar metals come into contact in damp environments. Structural steelwork should be painted with a suitable primer before as well as after assembly to prevent moisture seeping into the joints, causing rusting.

10.11 General rules for assembly and dismantling

The following rules should be observed when assembling machines and plant from new and when reassembling after maintenance and repair:

(a) The operation of the assembly must always be borne in mind. For instance, when the carriage is assembled on to the bed of a lathe, it must slide freely along the full length of the bed ways without binding at any point, but at the same time should show no slackness which could lead to loss of accuracy.
(b) The assembly drawings and specifications must be followed if they are available.
(c) The sequence of operations must always be planned prior to commencing assembly.
(d) All the parts and tools required must be available, clean and in good order.
(e) The penetration of dirt and other undesirable substances must be prevented during assembly.
(f) The correct fasteners must be used for all joints.
(g) Tightening tools must be of the correct size and type and fit properly.
(h) Parts must be assembled in accordance with any identification marks (Figure 10.3) or instructions.
(i) Checking and inspection should be carried out, as often as the work permits, in terms of position, clearance, truth, dimensions and operation.

The following rules should be observed when dismantling machines, plant and systems:

(a) Assembly drawings and specifications must be followed (in reverse) if they are available. Otherwise notes and sketches must be made to ease reassembly. Sometimes special dismantling instructions are available.
(b) Always plan the sequence of operations before dismantling.
(c) Select the correct tools and equipment.
(d) Imperfection such as burrs and corrosion should be removed to ease the dismantling process or avoid further damage.
(e) Particular attention should be paid to the selection of the correct tools if hammering is required during dismantling.

Note the importance of observing correct safety practices during assembly and dismantling, especially where large and heavy equipment, flammable gases, dangerous chemicals, steam or compressed air, or electrical services are involved.

10.12 Assembling pipework systems

Systems carrying fluids must be carefully planned to ensure that the pipework fittings and installations are suitable for the particular service under consideration. Pipework for the removal of waste and effluents requires a totally different approach to the hydraulic system of an earth-moving machine.

The factors which should be considered include:

Working pressure
Steady or pulsating flow
Temperature
Type of fluid being conducted (corrosion)
Safety, including legal requirements of inspection, maintenance and insurance
Cost.

Many fatal and serious accidents have occurred through failures in high-pressure and high-temperature installations and, because of this, stringent safety codes of practice have been drawn up which are enforceable at law. These requirements ensure almost complete safety to those working near such installations. All installations operating at high temperatures and pressures should be:

(a) Designed by expert engineers to meet the requirements of the Health and Safety at Work Act, the Factory Acts, and any special legislation appropriate to a particular type of installation, for example, bulk chemical plants and oil refineries
(b) Constructed only from equipment which has been tested and certified, and which is in accordance with the recommendations of the British Standards Institution
(c) Inspected by the appropriate authorities upon completion, for example, local authority engineers and inspectors and insurance surveyors
(d) Inspected at regular intervals thereafter (usually annually) by authorized persons such as insurance surveyors
(e) Colour coded to indicate the contents of the pipework.

There are many different types of pipe and tube used:

Copper This is solid-drawn tubing suitable for pressures up to about 150 bars after annealing (soft) or up to about 350 bars as drawn (half hard) at room temperatures. The safe working pressure falls off as the temperature increases. Copper tubes and pipes are corrosion resistant and suitable for low- and medium-pressure fluids such as water,

hydraulic oil and compressed air. They are widely used for domestic water and gas supplies.

Steel: hot finished seamless This is a solid-drawn tube which can be used at the highest temperatures and pressures (approximately 600 bars), but it has a rough finish and cannot be used with compression-jointed fittings. It is used with screwed or welded joints.

Steel: cold finished seamless This is also a solid-drawn pipe, but it has a smooth surface and can be used with compression-jointed fittings. It can be used for the same services as hot finished, but is very much more costly.

Steel: electric resistance welded This is rolled from steel strip and seam welded. It is used for electrical conduit and low-pressure liquids. Resistance-welded stainless steel tubing is sometimes used instead of copper for domestic water and heating services, but it is more difficult to cut, bend and joint.

Cast iron This is used as large-diameter pipes for high-volume, low-pressure services such as waste and effluent disposal and bulk water supply mains.

Plastic Extruded plastic pipe is widely used for low-pressure services. Rigid and semirigid pipes and tubes are used for water supply and gas supply underground services, and for waste and effluent disposal. They are unaffected by water, oil and most corrosive substances. Flexible plastic tubing is used for low- and medium-pressure coolant, oil and compressed air services on machine tools. High-pressure hose used for hydraulic oil is reinforced by a high-tensile steel braid.

10.13 Pipe jointing

Screwed joints

These can only be used with relatively thick-walled steel tubes.

There are two systems of screwed joint in common use:

Conduit thread This is a fine constant-pitch parallel thread used on electrical conduit. Since the conduit often forms the earth fault return path, the joints must be tight and *no jointing compound*

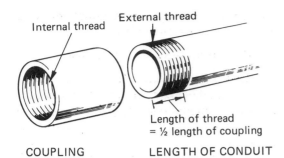

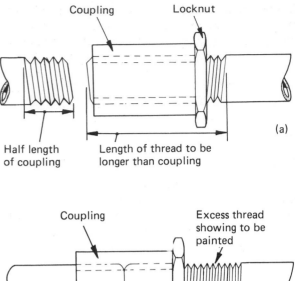

Figure 10.11 Jointing conduit for electric cables
preparing conduit:
1 Conduit is cut to length keeping end face square.
2 Bore of conduit should be reamed to remove any sharp edges. This is to prevent damage occurring to the wires when they are drawn in.
3 Cutting compound smeared on the outside of the tube and the stocks and dies used to cut the external thread. Dies should be reversed every few forward turns to break the cuttings and clear the thread.

Figure 10.12 Making a running joint:
1 External thread on one of the conduits is made longer than the coupling. The locknut and coupling are then screwed on to this conduit (first illustration)
2 The two lengths of conduit are then butted together and the coupling screwed back on the second conduit. When the coupling is up tight, the locknut is screwed up to the coupling to prevent it turning (second illustration).
3 Excess thread showing on first conduit is then painted to prevent rusting.

may be used as the joint must be electrically conductive. Figure 10.11 shows how conduit is joined and Figure 10.12 shows how to make a running joint.

British Standard pipe thread This is a fine constant-pitch system. The thread on the end of the pipe is tapered and the thread in the fitting is parallel. This enables the fitting to be pulled up tight on the pipe to ensure a fluid-tight joint. A jointing compouind is put on the threads before assembly, or PTFE tape can be wrapped round the threads.

Flanged joints

These are used for medium- and high-pressure joints which may have to be dismantled from time to time without disturbing the pipework installation, for example on pumps and valves.

There are two methods of attachment of the flanges to the pipes:

Screwed joints as described in the previous section.
Welding This is used for large-diameter pipes.

Whichever method is used, it is essential to ensure that the flanges are aligned axially and that they are parallel to each other and perpendicular to the pipe axis if a sound joint is to be made. The bolting holes must also be in alignment; they should be arranged so that no bolts lie on the vertical centre line, so that seepage will not cause corrosion.

Compression joints

These are increasingly used in place of screwed

joints for the smaller sizes of pipe. Not only do they save installation time, but pipe of lighter (and cheaper) gauge may be used since the wall is not reduced in thickness and weakened by threading.

Sleeved joints

These are used in conjunction with cast iron pipes, and a number of proprietary systems are available from the makers of the pipes.

Plastic pipes

There are a wide variety of jointing techniques used, including friction butt welding (gas mains), special compression joints for high-pressure hose fittings, and simple O-ring sealed joints which can be pushed together for waste pipes for wash basins etc.

A selection of these joints for pipes carrying fluids is shown in Figure 10.13.

Pipes, particularly those made of metal, expand when heated and contract when cooled. This has to be allowed for in long pipe runs. Special telescopic joints are available, as shown in Figure 10.14a. Alternatively, expansion loops can be included in the pipework at intervals along the run, as shown in Figure 10.14b. Horizontal pipe runs should also be supported on roller supports to allow for movement during expansion and contraction.

10.14 Dismantling pipework systems

Pipework has to be dismantled from time to time for replacement of failed components, replacement of corroded sections, removing blockages, and inspection.

The factors affecting the preparation for dismantling are:

(a) The shutdown time for the plant serviced by the pipework to be dismantled must be estimated.
(b) The correct operational sequence of shutdown must be observed, particularly when dangerous chemicals and high temperatures and pressures are involved (see Section 10.9).
(c) Prefabrication of replacement sections can reduce downtime.

(d) All tools, equipment and materials must be available before dismantling commences.

The shutdown procedures are as follows:

(a) Contents must be identified.
(b) Pipes must be depressurized and drained. Residual contents must be purged if dangerous. Expert advice and supervision is required.
(c) Machines must be rendered safe before work commences. For example, all machines must be isolated from their power source. Further, some components need to be jacked up so that they cannot fall: for example, the ram of a power press.
(d) Alternative routing of services may have to be made in plant which has to operate continuously.
(e) The safety procedures and codes for the plant must be strictly adhered to, not only on site but in the administrative preplanning and paperwork leading up to the dismantling programme. There must be notification of all workers, supervisors and managers concerned with the operation itself and those whose plant and departments are affected by the operation.

Pipework which has screwed joints and pipework employing male and female flanged joints present particular difficulties when dismantling if the surrounding pipework cannot be withdrawn as the joints are unfastened. As a last resort, the section of pipe concerned has to be destroyed by cutting out sufficient length to allow the joints to be withdrawn. No reclaimed materials may be used when reassembling. New bolts to the correct specification and new packing and gaskets must be provided. All joint faces must be cleaned of old jointing materials and checked for flatness.

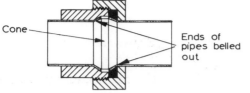

Cone

Ends of pipes belled out

(a) *COMPRESSION JOINTS*

(i) Manipulative joint suitable for soft copper pipes, etc.

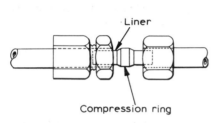

Compression rings

(ii) Non-manipulative joint suitable for copper pipes, etc., up to 50 mm bore.

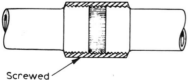

Liner

Compression ring

(iii) Suitable for plastic tubing. Note the liner which fits inside the two tubes, preventing them from being crushed.

(b) *SCREWED JOINTS*

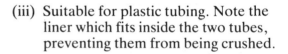

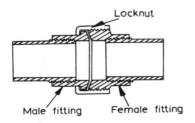

Screwed connector

(i) A binding material is usually wound into each of the male threads to effect the seal

Locknut

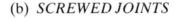

Male fitting   Female fitting

(ii) This type of fitting enables easy removal of portions of piping in a system. The joint is normally made with a jointing compound.

(c) *FLANGED JOINTS*

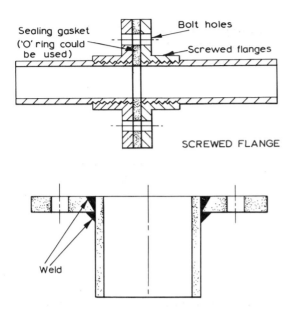

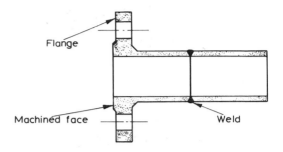

With the type of welded flange shown on the left, it is necessary to machine the face after welding to remove splatter and smooth the final run.

(d) *HOSE CONNECTORS*

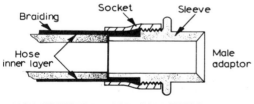

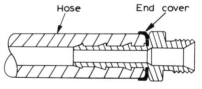

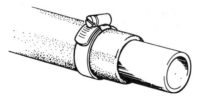

A worm-drive clip, of the type shown on the left, may be used to connect flexible hose to a metal pipe.

Figure 10.13 Joining pipes for fluids

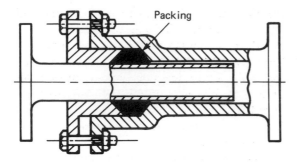

(a) EXPANSION COUPLING

Loop must lie
in horizontal
plane to avoid
vapour locks

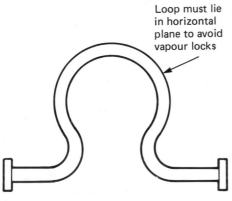

(b) EXPANSION BEND

Figure 10.14 Allowing for thermal expansion

Exercises

For each exercise, select *one* of the four alternatives.

1 Putting together a number of individual components to form a whole device is called
   (a) homogenizing
   (b) dismantling
   (c) assembly
   (d) structuring.

2 The correct sequence of assembly of the pulley on to the shaft as shown in Figure 10.15 is
   (a) 1,2,3
   (b) 2,3,1
   (c) 2,1,3
   (d) 3,2,1.

3 The dot punch marks in Figure 10.16 have been made by the fitter in order to
   (a) ensure correct positioning of any drilled holes
   (b) ensure correct interrelationship and positioning of the parts
   (c) indicate the sequence of assembly
   (d) caulk up a leak.

4 Components such as valves are often supplied with plastic covers sealing the holes. The main reason for these covers is to
   (a) protect the pipe joint seatings and prevent dirt entering the valve whilst it is in the stores

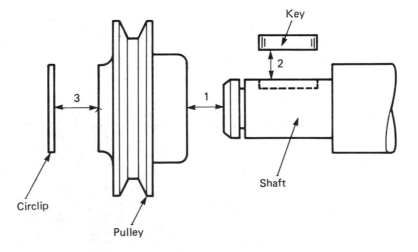

Figure 10.15

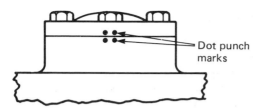

Figure 10.16

(b) protect the pipe joint seatings and prevent dirt entering the valve when it is in use

(c) prevent leaks occurring where the pipes join the valve

(d) provide inspection covers for maintenance.

5 In order to carry out periodic inspection and maintenance it is often necessary to
 (a) partially dismantle an assembly
 (b) carry out a trial assembly
 (c) totally dismantle an assembly
 (d) completely rebuild the equipment.

6 A distribution board carrying the switchgear, fuses and circuit breakers has been delivered to a factory ready assembled and only requiring to be connected to the machines using conduit and cables. This method of assembly is called
 (a) large-structure assembly
 (b) one-off assembly
 (c) on-site assembly: partial fabrication
 (d) trial assembly.

7 A machine gearbox is removed and sent back to the makers for repair. To avoid loss of production a reconditioned replacement gearbox is installed in place of the original. This is referred to as
 (a) on-site replacement and works reconditioning
 (b) on-site dismantling and workshop repair
 (c) one-off dismantling
 (d) on-site reconditioning

8 'Accessibility' when assembling and dismantling means that
 (a) the fitter has a permit to enter the working area
 (b) all parts of the assembly are easy to get at using standard tools

(c) the fitter has access to all the tools and parts required from the stores

(d) the designer is available for consultation.

9 Large structures are often subject to trial assembly before being dismantled into smaller units with the joints of adjacent sections numbered. This is to facilitate
 (a) transport to site and reassembly
 (b) reassembly only
 (c) working out the cost of assembly
 (d) training the assembly fitters.

10 When assembling large screwed fastenings it is *not* permissible to
 (a) extend spanners with tubing to get more leverage
 (b) use torque spanners
 (c) use ring spanners on hexagon head nuts and bolts
 (d) use soft steel washers under the nuts.

11 When a piece of equipment is reassembled after maintenance, all replacement parts must comply with the original specification
 (a) to ensure safe and reliable working
 (b) to avoid infringing patents
 (c) for ease of reassembly
 (d) so that they will fit the tools available.

12 The O-ring seals shown in Figure 10.17 are provided to
 (a) allow the piston to rotate in the cylinder whilst maintaining a fluid-tight seal
 (b) allow the piston to slide in the cylinder whilst maintaining a fluid-tight seal
 (c) prevent the metal piston touching the cylinder bore and causing wear
 (d) reduce the friction between the piston and the cylinder.

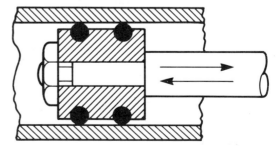

Figure 10.17

13 The device shown in Figure 10.18 is for
 (a) extracting wheels from shafts
 (b) pressing wheels on to shafts
 (c) straightening bent wheels
 (d) swelling the end of the shaft to fix the wheel in position.

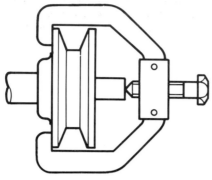

Figure 10.18

14 Penetrating oil is used mainly
 (a) when lubricating inaccessible parts
 (b) when oil pipes become blocked
 (c) when drilling deep holes
 (d) to assist the loosening of tight and rusted screwed fastenings.

15 Obstinate screwed fastenings can often be freed by heating the nut. This exploits the property of thermal
 (a) conduction
 (b) expansion
 (c) contraction
 (d) convection.

16 Nut splitters are used
 (a) only when dismantling structural steel-work
 (b) only when dismantling pipework
 (c) as a last resort when conventional methods of loosening screwed fastenings have failed
 (d) always when dismantling heavy-duty screwed fastenings.

17 Heavy components should be properly secured before dismantling commences so that
 (a) they can be easily lifted when dismantling is complete
 (b) they cannot be stolen when disconnected from the rest of the assembly

 (c) they cannot slip or fall and injure the maintenance fitter
 (d) faults can be more easily identified.

18 Pipes and conduits should be coloured
 (a) to brighten up the workshop
 (b) in accordance with BSI specifications so that their contents can be easily identified
 (c) so that any pipe can be used for any service just by changing the colour
 (d) so that no one will walk into them

19 To avoid accidental electric shock, electrical equipment should not be dismantled until
 (a) it has been earthed
 (b) smaller fuses have been temporarily fitted
 (c) a permit-to-work has been issued
 (d) it has been isolated from the supply.

20 Flanged pipe joints must *not* be fitted with the screwed fastenings on the vertical centre line because
 (a) this is the weakest position
 (b) fastenings in this position are inaccessible to fit
 (c) fastenings in this position are difficult to inspect
 (d) seepage from the joint may corrode the bottom fastening.

21 Before dismantling pipework it is essential that
 (a) the system is shut down and drained by the fitter
 (b) penetrating oil is applied to all screwed fastenings
 (c) the system is shut down, drained and purged by a suitably qualified and authorized person
 (d) old paint, corrosion and dirt is removed from the vicinity of the joints.

22 Expansion joints and bends are provided in pipework to allow for
 (a) expansion and contraction in length with changes in temperatures.
 (b) expansion and contraction in diameter with changes in temperature
 (c) extensions to the system
 (d) joints to be easily separated during dismantling.

23 When reassembling flanged joints,
 (a) always use new gaskets between the flanges

(b) always reuse the original gaskets

(c) only use a joint compound between the flange faces

(d) clean the flanges and scrape them flat so that no gaskets and jointing compounds are required.

24 When joining electrical conduit (metal type),

(a) never use a jointing compound or anything else which prevents metal-to-metal contact of the screw threads

(b) alway use a jointing compound on the screw threads so that water cannot get in

(c) always use PTFE tape between the threads

(d) always use an insulating compound in all joints.

25 Thin-walled copper tube for domestic services is best connected using

(a) screwed couplings

(b) compression or soldered joint fittings

(c) flanged and screwed couplings

(d) sleeved joints with O-ring seals.

Answers to exercises

Chapter 1
1 c
2 c
3 a
4 b
5 d
6 a
7 c
8 d
9 b
10 a
11 b
12 c

Chapter 2
1 c
2 b
3 a
4 c
5 d
6 c
7 a
8 b
9 a
10 c
11 b
12 d
13 c
14 c
15 b
16 a
17 b
18 b

Chapter 3
1 a
2 c
3 b
4 b

5 c
6 a
7 d
8 a
9 b
10 c
11 a
12 c
13 a
14 b
15 c
16 a
17 b

Chapter 4
1 b
2 a
3 b
4 d
5 a
6 b
7 c
8 c
9 a
10 b
11 d
12 b
13 b
14 c
15 d
16 c
17 d
18 a
19 b
20 b
21 c
22 a
23 a
24 b

25 b
26 d
27 c
28 a
29 d
30 a

Chapter 5
1 b
2 c
3 a
4 d
5 a
6 b
7 b
8 c
9 a
10 b
11 a
12 b
13 d
14 d
15 a

Chapter 6
1 b
2 c
3 a
4 c
5 d
6 d
7 a
8 b
9 a
10 c
11 a
12 a
13 b
14 d
15 c

16 d
17 c
18 a
19 d
20 c

Chapter 7
1 b
2 a
3 c
4 c
5 a
6 d
7 a
8 c
9 a
10 b
11 c
12 d
13 b
14 a
15 a
16 c
17 a
18 b
19 a
20 d
21 b
22 b
23 b
24 d

Chapter 8
1 d
2 c
3 a
4 d
5 a
6 b
7 d

8 a
9 a
10 d
11 b
12 d
13 a
14 b
15 c
16 d
17 c
18 b
19 a
20 d
21 c
22 b
23 c
24 d
25 a
26 b
27 b
28 a
29 b
30 d

Chapter 9
1 b
2 d
3 b
4 d
5 d
6 c
7 c
8 c
9 b
10 a
11 d
12 a
13 b
14 d

15 d
16 b
17 b
18 b

Chapter 10
1 c
2 c
3 b
4 a
5 a
6 c
7 a
8 b
9 a
10 a
11 a
12 b
13 a
14 d
15 b
16 c
17 c
18 b
19 d
20 d
21 c
22 a
23 a
24 a
25 b

YEOVIL COLLEGE

LIBRARY